WHAT READERS ARE SAYING ABOUT
A FRIDAY FILLED WITH JOY

✦ ✦ ✦ ✦ ✦

Menlo Innovations is an extraordinary company. And this is an extraordinary book. No theory. No abstractions. We learn the Menlo "secrets" by watching a day unfold. Essentially, a minute at a time. You develop a deep understanding of the Menlo Way courtesy of an approach that is compelling and entirely original. Bravo!

Tom Peters
Chief Provocateur, Tom Peters Company
Author of *In Search of Excellence* and *The Excellence Dividend*

✦ ✦ ✦ ✦ ✦

Organizational culture is the disruptive technology of our age. While conventional wisdom points to product innovation and the impact of artificial intelligence and a gig economy on the future of the workforce, sustainable success will not be determined by leaders who devalue employees. Attraction and retention of the best talent will be driven by organizations that place high value on respect, inclusion, transparency, development of knowledge and creativity, and promotion of a balanced lifestyle. In the end, the best talent will vote with their feet if they feel devalued and commoditized.

Michael Pacanowsky has dedicated the better part of his career to studying the elements of organizational culture that set great companies apart. His account of cultural differentiation at Menlo Innovations is a must read for anyone seeking to build sustainable success with top talent.

Tom Lewis
Founder, Noble 4 Advisors
Former CEO, The Green Exchange, Ameritrade Holdings, APX and Campus Pipeline
Former head of technology for the Executive Office of the President, The White House

✦ ✦ ✦ ✦ ✦

Organizational culture has become a trendy topic within the business world these days. While many companies strive to achieve the "perfect" culture by offering free snacks, open office space, or unlimited time off, few have studied the deeper meaning of culture. *A Friday Filled with Joy* offers the rare glimpse into a company that has done just that. This story, focused on a singular day at Menlo Innovations, narrates how an agile approach, intersected with "kindergarten skills" and aligned with the Thriving Organizational Culture model, can create a workplace culture full of joy. By following employees throughout their day, and even taking a quick journey back into Menlo's history, we see how culture isn't something we aspire to or simplify into a few buzz words. Rather, we begin to visualize how culture develops and evolves over time, trickling into every facet of how we work – interview practices, salary discussions, job titles and descriptions, meeting schedules, office layouts, development opportunities, and even work/life balance. Yet, what is truly remarkable about Menlo is no matter what challenges arise in their work, the culture has become so deep-rooted that it always brings these Menlonians back to what matters most – joy.

Annie Leither
Marketing Executive

✦ ✦ ✦ ✦ ✦

I have been immersed in agile/iterative methods my entire career and observed a wide range of what people consider to be agile/iterative methods and principles. Too often, we focus on the methods and not the underlying principles. This book is the most pragmatic guide for understanding the foundation principles and culture I have ever read—by far. This "day in the life" explains the culture that supports agile—transparency, trust in the individual and team, gaining context and understanding before making decisions and the willingness to experiment and learn—and supports that with specific, real examples of how to practice this culture and principles. This is the best how-to on agile that I have experienced. Each element of this "day in the life" taught me the ways I can demonstrate, and measure improved culture, principles and methods.

A Friday Filled with Joy is well-written, engaging and is now one of my go-to sources for myself and my teams as we seek to continuously improve. In other words, I love it!

Niel Nickolaisen
Senior VP and Chief Information Officer
O. C. Tanner Co.
Agile Culture (co-author)

✦ ✦ ✦ ✦ ✦

A Friday Filled with Joy is an exceptional view into a truly exceptional workspace, Menlo Innovations. The book is a true companion to Rich Sheridan's (CEO and Chief Storyteller of Menlo Innovations) books and brings new insights and appreciation not only for the day-to-day life at Menlo but also the heart and people that enable this great experiment. Reading this book, you quickly realize that the special sauce at Menlo isn't Rich and James, the Viking helmets, the special corkboard or tables on rollers – it is the respect everyone in the organization has for one another as fully rounded, complicated human beings. Michael and his team bring to life the idea that success can come from constantly striving to understand what drives people, whether they be customers, collaborators or teammates, and placing all of that in a supportive and caring environment that recognizes there is always opportunity to learn, grow and try new things.

Thomas Comery
VP, Strategy, Portfolio Operations and Communications, Biogen, Inc.

✦ ✦ ✦ ✦ ✦

A Friday Filled with Joy invites you into an intimate tour of the heartbeat of a great organization. Pacanowsky and team unlock the culture of Menlo Innovations as never before. This behind-the-scenes look at their daily rhythms is simultaneously inspirational and practical. It is a must read for those committed to Truly Human Leadership. A Masters Class case study of a truly inspirational organization, *A Friday Filled with Joy* empowers us all to see what is possible in business today and then go and make it happen in our organizations!

Brian Wellinghof
Director, Strategy, Improvement, Culture
Barry-Wehmiller

+ + + + +

When I read this book, what struck me was the intention to build deep relationships while instilling accountability and excellence. A few examples:

- The Daily Stand Up meetings are wonderful relationship building events that are in context and hold the team members accountable.
- The Pair Programming, although initially awkward, creates a deep sense of responsibility and respect between the two individuals.
- The Hiring process involves many team members and once the individual is brought on board it is the team's responsibility to help that person be successful. One of the most powerful parts of this process is the feedback on performance AND behavior in the first weeks of employment.
- The proximity of the work environment allows real time interactions and feedback in the moment which builds trust very quickly and enhances productivity.

The summary for me – At Menlo they have found a way to call forth the heart of a person to bring the Mission of the Company to life!

Thomas M. Carmazzi
Retired CEO of Tuthill Corporation

+ + + + +

In this day-in-the-life exposé, Michael and his team pull back the curtain on Menlo Innovations, revealing "Joy" is serious business. It gives a rare opportunity to see, not only what Menlonians do differently, but understand why they do it. In a single day, the team learns the backstories that connect Menlo's innovative practices to its foundational pursuit of Joy. What initially seemed counterintuitive was, in fact, Menlo's common-sense solution to the destructive customs so prevalent in other companies. With Menlo as an illustrative backdrop, Michael shares insights on his thoughtful and purpose-driven approach to developing a thriving organizational culture. To that end, A Friday *filled with* Joy is also filled with hope.

Douglas O. Harris, Ph.D.
Founder, Neuvante Solutions LLC

✦ ✦ ✦ ✦ ✦

Michael Pacanowsky and his team have adeptly opened the window into a high-performing culture in action, succeeding in their goal of helping the reader readily and easily absorb the day-to-day actions of Menlo Innovations. As I work on my own business, as well as with clients in their startups and operating companies, *A Friday Filled With Joy* is now one of my must-have reference books, a veritable "bible" for any culture-enhancing entrepreneur to reference.

If you are an entrepreneur, executive, or business owner, who cares deeply about creating and molding a business where people love to work and perform at a high level—while helping grow your company's revenues and increase profits—I encourage you to read *A Friday Filled With Joy*, and make it a part of your permanent library.

Bryce Hansen
Co-founder, Culinesco
Associate Director, Salt Lake Small Business Development Center

✦ ✦ ✦ ✦ ✦

Following the employees and founders of Menlo Innovations during a typical day, this book reveals what's possible when the fixed points in an organization's culture are identified and challenged. Examining the team's daily tasks and obstacles through the lens of the thriving organization model, your thinking will shift from 'That would never work here!' to 'We could do something like that!' No matter where you (or your team) sit within an organization, everyone has the power to run the experiment and drive change.

Mary Wheelwright
Senior Business Operations Manager
FireEye, Inc.

✦ ✦ ✦ ✦ ✦

In *A Friday Filled with Joy*, Pacanowsky and team give us a compelling view into a company known for building an innovative, resilient and successful culture. This book is an intriguing read for those who believe a workplace culture can be a place of joy. The insider perspective helps the reader understand what it feels like to be a part of this intriguing company.

Rachel Klemens
Director II, Culture and Engagement
CHG Healthcare

* * * * * *

Organizational culture is too often misunderstood and mismanaged. In this beautifully written book, Michael Pacanowsky and colleagues give us a rare glimpse into a day within a truly expectational culture. More importantly, they demonstrate how the stated values that form the bedrock of culture are brought to life through behavior and the organization's systems. A must read for those who want to understand, design and cultivate a thriving culture.

Reza Ahmadi, Ph.D.
Managing Principal, Emergent Solutions, Inc.

* * * * * *

A gem of a book about high-performing organizational culture – entertaining, accessible, and substantial!

Michael Pacanowsky and his team of honors college ethnographers share an intimate account of a day in the life of an extraordinary organization, Menlo Innovations. But by presenting their account through the lens of Pacanowsky's model of high-performing culture, developed in *The Thriving Organization*, Menlo's practices are revealed as carefully designed manifestations of Menlo values. Pacanowsky reminds us that culture lies beneath the daily routines – in principles, values, and axioms. Pacanowsky shows us why the practices and routines of Menlo produce high-performing culture. There's no checklist or step-by-step process for building culture, only the thoughtful, intentional and persistent application of values and principles. Congratulations to Rich Sheridan and James Goebel for what they have created at Menlo and thanks to Michael Pacanowsky for giving us deep understanding about what it takes to create joyful, thriving organizations.

Jerry Benson Ph.D.
President, 3rd Wind Leadership
Former CEO, Utah Transit Authority

A FRIDAY FILLED WITH JOY
One Day in the Life of a Radically Innovative Company

MICHAEL PACANOWSKY
SUSAN ARSHT
VICKI WHITING
SARA D'AGOSTINO
MAGGIE FISCHER
RACHEL IVERSON
ELIZABETH JOHNSON
COLE POLYCHRONIS

with
Jim McGovern

A Friday Filled with Joy
One Day in the Life of a Radically Innovative Company
All Rights Reserved.
Copyright © 2021 Michael Pacanowsky, Susan Arsht, Vicki Whiting, Sara D'Agostino, Maggie Fischer, Rachel Iverson, Elizabeth Johnson, Cole Polychronis, with Jim McGovern
v4.0

The opinions expressed in this manuscript are solely the opinions of the author and do not represent the opinions or thoughts of the publisher. The author has represented and warranted full ownership and/or legal right to publish all the materials in this book.

This book may not be reproduced, transmitted, or stored in whole or in part by any means, including graphic, electronic, or mechanical without the express written consent of the publisher except in the case of brief quotations embodied in critical articles and reviews.

Outskirts Press, Inc.
http://www.outskirtspress.com

ISBN: 978-1-9772-3010-2

Cover Photo © 2021 Michael Pacanowsky. All rights reserved - used with permission.

Cover Design by Sophia O'Brien and Summer Shumway.
Book Design by Sophia O'Brien.
Photographs used with permission of Menlo Innovations.

Outskirts Press and the "OP" logo are trademarks belonging to Outskirts Press, Inc.

PRINTED IN THE UNITED STATES OF AMERICA

DEDICATION

We dedicate this book with grateful affection
to our families.

We also dedicate this book with earnest hope for success to all like-minded and like-hearted folks who are working to create organizations that are, as Gary Hamel puts it, "fit for the future and fit for human beings."

TABLE OF CONTENTS

ACKNOWLEDGMENTS i

WHO SHOULD READ THIS BOOK AND WHY iii

HOW TO READ THIS BOOK iv

AN OUTSIDER'S FOREWORD, *Doug Kirkpatrick* vi

AN INSIDER'S FOREWORD, *Rich Sheridan* ix

FRIDAY 1

AFTER FRIDAY 207

EPILOGUE 212

ABOUT THE AUTHORS 252

APPENDIX 254

ACKNOWLEDGMENTS

We want to thank Rich Sheridan, James Goebel, and all the Menlonians who supported the idea of having a group of faculty and students from Westminster College spend a full day observing them and then publishing in intimate detail what we observed. Menlo has been open to visits by many people over the years—academics, business leaders, students, members of the business press—and many people have written about what they saw. But we were the first group to descend on Menlo with so many for so long and to write so much. We also thank Anna Boonstra, who led the Menlo Experience team, for her help in arranging the specific Menlonians that we shadowed, and in handling the logistics of our visit. Lisa H provided valuable assistance in post-research visits and interviews to round out our understanding of Menlo ways.

We also want to thank Richard Badenhausen, the Dean of the Honors College at Westminster, who was sufficiently intrigued by the idea of having Honors students participate actively in this project that he helped identify potential candidates. We thank Orn Bodvarsson, Dean of the Bill and Vieve Gore School of Business at Westminster, for his support for the Center for Innovative Cultures. And we thank the Center's team—Michael Zavell, Summer Dawn Shumway, and Sarah Hirst—who held the fort down while we were away in Ann Arbor or were squirreled up writing.

David Noller, Jerry Benson, and Miriam Eatchel all read an earlier version of this book and made valuable suggestions. We particularly want to recognize Miriam for her suggested subtitle of this book, and her insights into what this book was "really" about! David Noller, a Hollywood screenwriter, encouraged us to scrap a lot of the data we had incorporated in an earlier version of the book, drawing on the analogy of "making a better film by leaving lots of pieces of it on the cutting-room floor." He also sparked the idea of writing an Epilogue that outlined the perspective of high-performing organizational cultures that we took into our research on Menlo Innovations.

Finally, we want to thank the donors of the Center who made it possible for us to undertake this research. We especially want to thank our founding donor—Ginger Giovale. Ginger has been a steadfast supporter of Westminster College and the Center for Innovative Cultures. Thanks, Ginger! We also want to acknowledge timely support from Judy and Bing Fang who helped make turning our research into a book possible.

Note: The photos in this book are photos provided by Menlo Innovations and used with their permission. No photos were taken the day of our research. These photos, then, are representative of various Menlo events and activities but not the specific ones described in this book.

Also note: At the request of some of the Menlonians who are featured in this book, we've changed names for reasons of privacy. Also, for client confidentiality, we've changed the names of client team members and the names of the projects that Menlo was working on for them, as well as modified the actual nature of some of these projects.

Richard Sheridan and James Goebel are the actual names of Menlo's co-founders.

WHO SHOULD READ THIS BOOK AND WHY

Gary Hamel, the management guru, claims that the key challenge we face in organizations today is our need to "create organizations that are fit for the future and fit for human beings." Given the current speed of change, the levels of uncertainty that organizations face, the new generations of workers who want something like fulfillment and not just a paycheck from their work, our legacy management thinking and practices from the 20th century just won't cut it.

Menlo Innovations is an organization that demonstrates what it's like to be fit for the future and fit for human beings.

If you are a front-line employee, or maybe a person of some influence in your organization—a supervisor, manager, division head, CEO—or just someone who is interested in interesting companies and how they work, learning about Menlo should:

- Inspire you to imagine new practices you might take to your own organization.
- Give you the confidence that these ideas can actually be implemented in, and adapted to, your organization.
- Provide you with a deeper understanding of how organizational culture works and why understanding it is so crucial to the successful introduction of positive change in your organization.

HOW TO READ THIS BOOK

This book is mostly about Menlo Innovations. It takes you through a day at Menlo, a day that is somewhat ordinary, because it is typical of what happens on any given day, and extraordinary, because it allows a deeper explanation of why Menlo's people are able to do such amazingly radical and innovative things!

This book is a little bit about the software development industry and a relatively new approach to software development known as Agile. Because some readers may not be familiar with Menlo, software development, or Agile, we've inserted enough information on the subject that you shouldn't have any trouble understanding some of the technical aspects of what's happening at Menlo on this particular Friday.

This book is also quite a bit about organizational culture, and a particular perspective on organizational culture that we have developed at the Center for Innovative Cultures at Westminster College. We believe that what happens at Menlo, and why it works, has to do with the company's underlying culture and how *it* works. Some of our ideas about organizational culture, including Menlo's organizational culture, are highlighted throughout the book. The Epilogue at the end of the book takes a deeper dive into our perspective on organizational culture, and again, how Menlo's culture actually operates and on what it's based.

As you read this book and come across things that Menlo does that you find interesting and maybe worth emulating, ask yourself these questions:

- *Why does Menlo work so well? Why is this small company in Ann Arbor, Michigan of such interest to so many people worldwide?*

It isn't because Menlo has some proprietary technology that gives them a competitive advantage no one else can copy. It isn't because Menlo people all take a magic pill each morning before they come to work that makes them productive, collaborative, and (dare we say it) joyful. What makes what Menlo does possible? (Hint: It's their culture.)

- Think of some other organizations you're familiar with, maybe but not necessarily your own. Could these organizations introduce any of these Menlo practices? Why or why not? There may well be good reasons why a direct copying of a Menlo "best practice" just won't work. But would something similar or analogous work? Why or why not? (Hint: It might be the culture.)

AN OUTSIDER'S FOREWORD

Doug Kirkpatrick

The Global Drucker Forum in early November 2015 at the Hall of Sciences in Vienna was a special event in a special place. It was my first Drucker Forum and a chance to meet friends and hear from several of my business thought-leader heroes: Tom Davenport, JC Spender, Henry Mintzberg, Rita Gunther McGrath, Bill Fischer, Charles Handy, John Hagel III, Zhang Ruimin, and Steve Denning.

I had a passing familiarity with the story of software house Menlo Innovations through my association with the Center for Innovative Cultures at Salt Lake City's Westminster College and its tireless head, my friend Michael Pacanowsky. Through my work with Morning Star and other vanguard companies, I had a burning passion for unlocking the secrets of human-centric workplaces. Rich Sheridan's book *Joy, Inc.: How We Built a Workplace People Love* was attracting attention around the world as a case study in creating a great company where people could flourish. The *Learning Consortium for the Creative Economy* (a group of eleven companies, including Menlo Innovations, working under the auspices of Scrum Alliance®) would be presenting a report on the future of work from Steve Denning, Jay Goldstein and Michael Pacanowsky. Electricity was in the air as the event got underway. It was an exciting time to network, listen and learn. Thanks to Michael's thoughtful connection-making, one of the first people I met was Rich Sheridan.

The story of Menlo Innovations deserves wide circulation. Aiming for engagement is one thing; aiming for joy is something else altogether. The story of creating a workplace culture where people combine high performance with care, compassion and wholeness is compelling.

Diving into that culture for a day (one day!) to take a culture core sample to see if reality matches publicity is simply brilliant. Leveraging the themes of Sheridan's first book, *A Friday Filled with Joy: One Day in the Life of a Radically Innovative Company* by Michael Pacanowsky, Susan Arsht, Vicki Whiting, Sara D'Agostino, Maggie Fischer, Rachel Iverson, Elizabeth Johnson and Cole Polychronis highlights Menlo Innovations (and its Menlonians) in a captivating way. Co-authors Susan Arsht and Vicki Whiting (whom I also personally know to be first-rate organizational thinkers) assuredly contributed greatly to the data-gathering and structure of this thoroughly engrossing book.

One of Menlo Innovations' unique features is the use of High-Tech Anthropologists (HTAs) to work with primary users to derive the real story of what users really need and want in their application software to get the job done. When I met Menlo anthropologist Mollie Callahan at an event in Salt Lake City and she shared her job title, I pictured her studying a stone-age tribe in a distant jungle à la Margaret Mead. After Mollie explained a bit about what she actually did at Menlo, the picture started to clear up. Now, having read several vignettes about the joy-filled Friday at a radically innovative workplace, her role as a High-Tech Anthropologist is more than clear. It's really about delighting the customer by knowing the customer's users as well or better than the customer knows its own users (in Menlo-speak, the "primary personas"). This approach directly supports the Mission of Menlo Innovations: to end suffering in the world as it relates to technology.

Other vignettes followed various development teams (they always work in pairs, a productivity, quality and teamwork innovation from Sheridan's days with Interface Systems); QAs (Quality Advocates), Project Managers, Menlo co-founder James Goebel, various teams, weekly Show & Tell sessions for customers, the Daily Stand Up (inclusive of the entire company, which virtually always takes eleven minutes to complete); and Rich Sheridan himself as he attends various meetings and rides in an autonomous vehicle at a test facility as part of his community outreach.

Community kitchen care, rituals, stories, storycards, audition interviewing, open workspace, high speed voice technology (calling out verbally to anyone or everyone in the company), Viking helmets (real Vikings, not Minnesota Vikings), Eeyore dolls, open books (including pay rates), babies at work, dogs at work, profound-yet-pithy quotations posted on the walls, a disco ball, "Do you have time for this?" and other artifacts and processes describe a deeply human culture fully capable of promoting joy. Once I started reading, I couldn't put the book down—it was like being a proverbial fly on the wall, completely fascinated by the details. The authentic "AFTER FRIDAY" follow-up was a nice touch, reminiscent of an on-screen movie epilogue. Michael Pacanowsky's substantive and thoughtful coda provides an excellent summary of distilled wisdom from his deep dive into Menlo Innovations and his career turns with W.L. Gore & Associates and the Center for Innovative Cultures.

Culture is a tricky thing. Every organization has a culture. One can choose to be intentional about culture, or default to whatever culture happens to emerge. Who wouldn't want to create a culture that people love, that brings joy, that ends suffering? Menlo Innovations has an approach that works for them. Try to superficially copy them and court failure. Learn their deeper lessons and embark on a journey of discovery and joy.

Doug Kirkpatrick
Author of *The No-Limits Enterprise:*
Organizational Self-Management in the New World of Work
San Francisco, California
May 24, 2020

AN INSIDER'S FOREWORD

Richard Sheridan

The Dream of a College Student

I've always felt the software industry, my professional home since high school in the early 70's, attracted more entrepreneurs per capita than almost any other field. Of course, my unscientific data collection could be evidence of confirmation bias. All I know is that I wanted the thrill of having my own company since I was 14 years old and it was going to be a software company. I had never imagined it would take until I was 43 years old for that dream to come true.

Though the wait was long, the thought never left me.

In the early 80's as a college student at Michigan (naturally Computer Science was my chosen academic pursuit), I was daydreaming about my own future as I walked along State Street near Nickels Arcade, less than a block from the office we currently occupy. I dreamt of my own company, one filled with life, with human energy, with camaraderie, with progress, serious technical breakthroughs, where people would work together in close collaboration to create something meaningful and world changing. We would be proud of our work and we would be noticed for what we did. I had done enough programming by then to also recognize that the means of that software creation had to be thoughtful too. Software could be too fragile otherwise, and there would be no pride in that. We would be VERY good at what we did, and we would be VERY good at *how* we did it.

I tucked that dream away in a box and put it up in the attic of my brain where it gathered dust for almost 30 years. That box waited for me, like a gift, to be opened and the unwrapping would be a breathtaking reminder of how powerful the dreams of youth can be.

Dream Interrupted by Life

I married, graduated, found my first job, and began life as a full-fledged adult with responsibilities and the seemingly inevitable troubles of work. A house, three beautiful daughters, two cars, my wife Carol staying home to raise our girls, formed the daily yoke of a young dad and husband. The jobs that followed turned into sour experiences as chaos and bureaucracy prevailed. I never had a chance, as W. Edwards Deming would say, "to work with pride." I'd come home tired, upset about either real problems or political problems at work. Often, my wife would ask me "honey, you look tired. Did you get a lot done today?"

No.

I was getting nothing done. I was busy, ran from meeting to meeting, often feeling like a firefighter and arsonist at the same time.

Something was gonna give, and I was determined that it wasn't going to be me. I knew there had to be a better way. A better way to organize the people, the process, and in doing so, get to pride and get to joy.

Never Give Up

My disillusionment fueled my passion. I became a student again, learning mostly from authors like Tom Peters, Peter Drucker, Peter Senge and John Naisbitt. I could see, in the fog of my advancing career, that there was indeed a better way. Nobody was writing "do this" and it will all be better. What they did say, "don't give up on your dreams! Others have succeeded, you can too!"

That was all I needed to keep hope alive.

The weirdest part of this entire journey was the success I was achieving by worldly measures. I had everything: job, title, authority, pay, stock options worth millions, and the respect of the people I worked for and those who worked for me. What else was there? Later, I would learn to clearly describe what I was missing … and I could boil it down to one word … **joy**. I didn't have it.

"How dare you!" I thought. You need joy too? I've come to realize it's all we need and everything else we chase is a poor substitute … *empty career calories.*

As good fortune would have it, I met, and engaged, an equally frustrated and passionate consultant named Clement James Goebel III. James would become my pair partner in a transformation that ultimately would garner world-wide attention. Over two years we would work tirelessly together to experiment our way to the transformation of a tired old public company on the west side of Ann Arbor called Interface Systems, Inc. where I was the VP of R&D. It worked! The stock rocketed from $2/share to $80 almost overnight. We did it. We had married business success with joy. The authors that encouraged me were right. You CAN have it all.

Then. It. Was. Gone.

In an instant.

The dot-com bubble burst and threw us all out on the sidewalk. No job, no title, no authority, no paycheck, no office, no team, stock options worth $0. Everything was gone except one thing: what James and I had learned in those two years. From the rubble of 2001's bubble burst, we became entrepreneurs and launched Menlo Innovations on June 12, 2001. By May 2003, we were a Forbes cover story. We quickly decided on a mission to "end human suffering in the world as it relates to technology" and establish a goal of returning **joy** to one of the most unique endeavors mankind has ever undertaken ... the invention of software.

Yes, joy.

The Long-Forgotten Gift Is Reopened

The work of entrepreneurship is a labor of love. Long hours, lots of learning, scary moments, and the joys of success ... we've had them all. We never gave up ... not during 9/11, two wars, the Great Recession, or even as I write this, the Great Pandemic. Every once in a while, as James reminds me, we need to stop climbing the mountain, take a breath, turn around and look back down to remind ourselves of how far we've come. That vista can be breathtaking. Then, turning back around, we get back to climbing.

One of those breathtaking moments happened for me by "accident." (I guess). I walked into Menlo, in our beautiful second office location on the third floor of the century-old Godfrey building in the Kerrytown district of old Ann Arbor. The sunlight was streaming in through the skylights, accentuating the old brick loft that had become our home. The Menlo Software Factory team was at work and fully engaged in our big open and collaborative space. I could see it, feel it, sense it ... the human energy, the laughter, the noise of work, the camaraderie, the Menlonians working shoulder-to-shoulder ... just as I had imagined it over three decades earlier.

I was hit with goosebumps then (and even now while writing this). I had the feeling of a Jack-in-the-box popping out of the gift-wrapped box in the dusty attic room of my mind saying "congratulations, you did it, you achieved the dream of your 20-year-old self." It hit me like a hurricane of emotion. How could this be? How is it possible that the never-revisited dream from 1980 could manifest itself so perfectly nearly thirty years later?

Pure joy.

My encouragement to you is this ... you too have hung onto those dreams. There is no reason to believe that you can't achieve them. These dreams can be your destiny.

The World Comes to See Us

Throughout our now nearly twenty-year company history, people have come to see us. They come from all over the world, now by the thousands each year. They want to see what joy looks like. They stay, they ask questions, they leave inspired to pursue their own version of joy.

One of those visitors was Michael Pacanowsky, the Director of the Center for Innovative Cultures at Westminster College in Salt Lake City. I don't know what Michael was expecting. In this visit, he was a volunteer for an industry expedition to study and learn from organizations who were seen as having "agile" cultures. I think it's fair to say that what Michael saw changed him. I was personally intrigued with Michael as he had come out of W.L. Gore & Associates, a company that I had admired from afar for years. Gore was an example of a company that achieved their version of joy AND business success. That someone from Gore would admire us was too delicious to ignore. We could learn from each other. As time went on, James, I and so many Menlonians embraced the Center at Westminster and all it had to offer. We are all better off for this relationship.

See What They Saw

Never one to shy away from different ideas, Michael approached me. What if he brought in a team, professors and students from Westminster, to turn the focus of their cultural microscope up close and study Menlo for a day and write up what they saw in drippy detail. It sounded just crazy enough to try. ☺

The result of that day long study is here in your hands. In the pages that follow you will see Menlo in all of its detail. I'm never sure quite how to describe exactly what makes Menlo magical. You will get a chance to see what I saw the day my long ignored "gift of a youthful dream" was reopened. You may see things in these pages that I don't see anymore because I'm so close to it … because it's so much a part of who I am, and quite frankly it all seems normal to me now.

It's not normal. Menlo is still so special.

My joy has been back for twenty years and I will never be the same. My hope is that the work of Michael and his wonderful team of Westminster collaborators, students and professors alike, can bring you joy in the reading, but, beyond that, a lasting joyful pursuit that perhaps has been long ignored inside of you.

Welcome to *A Friday Filled with Joy*. We are all glad you're here.

6:45 AM
RICH SHERIDAN
CEO and Chief Storyteller

On this Friday morning, Rich Sheridan is the first to arrive at the Menlo Innovations "software factory" in Ann Arbor, Michigan. He flips on the lights, then weaves his way through the maze of tables where 50 or so "Menlonians" will soon be tackling the day's work. The way work gets done at Menlo, this particular arrangement of tables, the pods they form, even the make-up of the teams who share them, will be different next week. At Menlo, these things change all the time.

> Your world is PERFECTLY organized to create the BEHAVIOR you are currently experiencing.
>
> — VitalSmarts

A poster on the wall at Menlo Innovations

In the kitchen area, Rich empties the dishwasher, and makes the first pot of coffee—prepping the office for the beginning of the day. Depending on his speaking schedule, Rich travels about 40-60% of the time, so he is often out of the office. When he's in town, he likes the idea of getting the office ready for the other Menlonians.

Rich doesn't have an office at Menlo. No one at Menlo has an office. So, after getting the coffee going, he walks over to a nearby table to check his e-mail. He's a big man in his early 60s, with a "casual presence." A self-described eternal optimist, a description shared by those who've known him for any length of time, he's also a natural storyteller. Often, when reflecting on some new problem or challenge, he will suddenly launch into a story that, as it unfolds, speaks to the issue in an interesting way. Inevitably, when that happens, his face breaks into a broad smile.

Actually, Rich smiles a lot, which is probably not surprising for someone whose two books about his company are titled *Joy, Inc.: How We Built a Workplace People Love*, and *Chief Joy Officer*. Someone who has built a company that says of itself:

> At Menlo, we do more than design and build great software. Not that great software is a small thing. It's rare. But we aim for something higher. Our processes, our culture, our work ethic—they all aim toward a single goal: joy.

✦ ✦ ✦ ✦ ✦

The history of Menlo Innovations is a story of love, love lost, and love found once again. Rich Sheridan was in high school when computers first hit the market, and for him it was love at first sight. He got a degree in computer science at the University of Michigan and, still in love, went to work as a software developer. He did well, kept getting promoted, getting bigger offices, bigger paychecks, bigger bonuses, until one day, when he was an Executive VP at a company called Interface Systems, he suddenly realized that the love was gone.

What was happening to Rich was happening, and still happens, to many others in software development. There's too much stress, too many projects that run behind schedule, over budget, or both—and too often end up getting canned before they're even finished. And worst of all, there are too many projects where what the customer eventually gets doesn't do what they need it to do or doesn't work the way it's supposed to work.

Rich explains it this way. "The technology was fun. I started my career writing code. It's fun. It's a craft. It's like writing music. It's noble." Here he gets serious. "But imagine you run a bakery and throw away everything you baked. That characterized my early career."

Here he was enjoying all the outward trappings of success, but the passion was all gone. Looking to recapture the fun, the joy, the love, Rich began thinking hard about the kinds of organizational structures that might make for a more satisfying, and more productive, work environment. His search took him to books like Peters and Waterman's *in Search of Excellence;* Peter Senge's *The Fifth Discipline;* and almost anything by Peter Drucker. And he began a series of deep conversations with James Goebel, a programming consultant he had engaged. James was convinced, and he convinced Rich, that the problem with software development had little to do with technology and a great deal to do with people and the organizations in which they worked.

Influenced by some early thinking by proponents of "agile" software development (more about this in a bit), Rich and James decided to attack their software ennui by building an experimental skunkworks within the traditional structure of Interface Systems where developers could try out a new way to do their work. They found some unused open space, crammed in tables and chairs, computers and keyboards, power cords and cables, and called it the Java Factory.

Then they asked for volunteers. The programmers who signed up sat at moveable tables, and they worked in pairs, sharing one computer, with the pairs changing every week or two—a radical break with the traditional "lone wolf developer" working away in his own cubicle with earphones on, oblivious to everyone around him. But much to their surprise, many developers found they actually liked working this way, and customers were delighted with the results. And for Rich, programming started to be fun again.

As the story of the Java Factory made its way through the Ann Arbor computer community, people started coming to see what it was all about. Eventually, a California company called Tumbleweed was so intrigued that it bought Interface Systems, and along with it, the Java Factory.

Then the dot-com bubble caused it all to come crashing down.

With Tumbleweed looking to save money everywhere it could, purchase orders for critical new equipment were denied, and then finally, Rich got a note to call his boss in California: "Rich, we're closing down the Ann Arbor office. You have two weeks until we get all the packages put together. Don't tell anyone."

What was he going to do? At age 43, Rich could see himself entering the most expensive phase of his life. He had three daughters, one heading off to college, one halfway through high school, and the last just three years back. And after college, somewhere down the line, there would be weddings! So even though it had long been a dream, he wasn't thinking just then about starting his own business.

But then again, the work he had been doing with James for the last two years at Interface Systems had brought back the joy of technology for him. And he had no desire to move his family to Silicon Valley. And so, with all this rattling around in his head, one night a few days after the call with his boss Rich sat bolt upright in bed at 3 o'clock in the morning. He knew what he wanted to do! His wife, Carol, was startled awake, and when he told her he wanted to start a company with James, she said simply, "Go back to sleep, Rich."

But the idea wouldn't go away. As it turned out, James' consulting company was also getting clobbered by the collapsing IT industry, and he too had been reinvigorated by the success of the Java Factory, so he was game. As crazy as the idea might have seemed at first, Menlo Innovations was born.

The Menlo "Factory"

MENLO INNOVATIONS

Menlo Innovations is a custom software development shop. While it continues to grow, with 50 or so employees Menlo is still not a big company. They work across a wide range of industries, in some cases fixing or upgrading legacy systems, in others designing brand new apps. Some of their projects take months to complete, others take only a week or two. Some of their clients—like Ford, Nationwide Insurance, Walmart, and Bectin Dickenson—have a national and even global presence, but others are small local companies you have probably never heard of, unless you live in Ann Arbor, companies like Hygieia, A2Bikeshare, or Parabricks.

And yet, Menlo "punches way above its weight." Rich has been on the cover of *Forbes* and Menlo Innovations has been listed on *Forbes* "America's Best Small Companies." *Inc.* has called it one of 25 small companies that will change the world. And since 2013, when they first started keeping track, Menlo has held more than 300 workshops onsite, hosted more than 600 tours of their facility, and had more than 20,000 visitors—all of whom are very interested in just what is going on inside the "Menlo Factory."

People from the software industry are curious about how Menlo has successfully implemented agile development, when so many other companies have tried and failed. But Menlo also attracts interest from outside the software industry, from business leaders who want to know how Menlo has built an organizational culture that generates such a depth of commitment and engagement from its employees. How does the Menlo culture enable its people to not only deliver exceptional performance, but to experience real joy in the process?

This book is intended to shed some light on that absolutely critical question.

8:00 AM
MARILYN MacEWAN & McTAVISH
Office Assistant and office dog

Marilyn MacEwan was the second person into the Menlo office on this Friday morning. A 13-year Menlonian veteran, Marilyn is Menlo's general office assistant and unofficial culture keeper. She's also the owner of McTavish, the office dog.

Every workday morning, Marilyn makes the 90-minute drive from her home to downtown Ann Arbor. Every day, when she exits the interstate onto the surface streets that lead to the Menlo Factory, McTavish starts whining with anticipation. He's made this journey every workday of the last ten years, since Marilyn adopted him from a shelter, and he clearly looks forward to it.

Marilyn could easily find a place to work closer to her home, but her loyalty to Menlo runs deep. "They care about you as an individual and they understand that you have a life outside of work. If a child is sick, if you are sick, they take care of you. We feel like family."

By way of example, Marilyn tells this story. One day she woke up definitely feeling sick, but she went to work anyway. Later in the morning she had a meeting with Rich Sheridan, who quickly realized that she was not feeling well and told her to go home. As Marilyn explains, lots of bosses might have done that, but Rich took it a step further. He called her sister and had her take care of Marilyn when she got home. He stressed that the sister—he called her "the Enforcer"—was to make sure Marilyn was fully well before returning to work. As it turned out, Marilyn had pneumonia.

That deep-rooted caring for the well-being of all the people who are part of the Menlo family sits at the very core of the Menlo culture. To understand Menlo's success, we need to understand that culture.

THE CULTURE OF A THRIVING ORGANIZATION

What do we actually mean by an organization's culture? We know it has something to do with what people at work do, and how and why they do it. It's behavior and norms and values and understandings. But what's the relationship between an organization's culture and its performance? Are certain organizational cultures more conducive to high levels of employee engagement, to more effective and agile decision making, to greater customer satisfaction and loyalty, to a better bottom line? Is there, in fact, such a thing as a "high-performance" culture? If so, what does it involve, and how do you build one?

As the pace of innovation has accelerated and the need for organizational agility and resilience has increased accordingly, as new generations have brought new goals and values into the workplace, these questions have taken on greater and greater urgency. For the past several years, we at the Center for Innovative Cultures have wrestled with these questions. We've looked at the research and drawn on the insights of many leading thinkers. (We owe a particular debt to Edgar Schein, whose pioneering work introduced the concept of organizational culture and established a foundation for our work and that of many others.) And we've talked with hundreds of executives and visited an array of very diverse organizations in Europe as well as the US.

At least some of the results of this work are laid out in *The Thriving Organization: An Exploration of the Deep Dynamics of High-Performing Organizational Cultures*. That book draws on examples from many different organizations, including in-depth case studies of three organizations, one of which is Menlo Innovations.

Basically, the central argument of *The Thriving Organization* is as follows:

1. As per Schein, an organization's culture is a unique set of ways of perceiving, feeling, thinking, and acting—learned and consciously or unconsciously adopted in response to how the organization has met, or failed to meet, a set of internal and external challenges.

2. An organization's culture is often thought of as an array of specific behaviors, processes, and "artifacts"—its employee perks (foosball tables, meditation rooms, gourmet snacks, day care center), its dress code, preferred modes of communication, hiring practices, etc. A more rigorous definition, however, includes a set of interdependent axioms, values, and principles—underlying layers that build on one another and ultimately determine the more visible aspects of the culture.

3. Thriving organizations are those that achieve their strategic objectives while enabling (and in our view, *because* they enable) their people to be and *become* their best selves at work. Such organizations may look very different at the level of cultural practices and artifacts, but they are very likely to share similar axioms, values, and principles.

4. The **axioms** are the foundational level. They include: *Treat individuals with respect. Promote the common good. Make decisions at the lowest level. Focus on long-term success.* On top of the axioms is a set of **values** that includes *trustworthiness, caring, self-responsibility,* and *magnanimity.* Finally, along with these axioms and values, thriving organizations share a set of **principles** that include: *Connect the dots. Cultivate sensitivity to the outside world. Enable collaborative emergence. Invest in future (adaptive) capacity. Expect leadership everywhere.*

These cultural elements will be called out throughout this book by little sidebar boxes that will be signposts alerting you about what's going on beneath the surface of what you see or hear at Menlo. How these cultural elements fit together are examined in detail in *The Thriving Organization*, and in somewhat less detail in the Epilogue at the end of this book. (If you think it might be helpful to gain a deeper understanding of our perspective on organizational culture, including how it derives from and differs from Schein's model, feel free to read the Epilogue first.)

This book is intended to deepen that discussion of thriving organizational culture by showing at an even more granular level what such a culture looks like and how it manifests itself in the varied, often mundane activities of the 50 or so people who come to work every day at Menlo. To put it in different terms, this is what joy looks like on one ordinary Friday in one thriving organization.

8:05 AM
RICH SHERIDAN
CEO and Chief Storyteller

> ## CULTIVATE A SENSITIVITY TO THE OUTSIDE WORLD
> Cultivating a sensitivity to the outside world is more than just paying attention to current customers. It's about learning from potential future customers.

The first major item on Rich Sheridan's agenda this morning is the SPARK CEO Round Table. SPARK is the economic development agency of the Ann Arbor area. When Rich and James and their other two partners founded Menlo in 2001, Rich was somewhat shocked to realize that, although he had grown up in Ann Arbor, attended the University of Michigan, worked for various Ann Arbor-based computer companies for 20 plus years, he didn't really know the Ann Arbor community or business scene. So, having committed to the cultural axiom of *Promote the Common Good* when they founded the organization, Menlo was going to be firmly rooted in its community. With that in mind, Rich took on the role of being the company's "public face." In that role, involvement with SPARK is one of his responsibilities.

The CEO Round Table is held at a different location each quarter. About 20-25 people typically show up. Most are CEOs of local businesses. A few are from the University of Michigan or various local government agencies. This time the venue is the American Mobility Center adjacent to the old Willow Run airport. During World War II, Willow Run was the site of a 5-million square foot factory built by Henry Ford to produce B-24 bombers. The largest single-story structure ever built, the plant eventually produced 25 planes a day, an incredible feat of industrial might.

After the war, the Willow Run facility became a GM transmission assembly site. In 2008, when GM entered bankruptcy in the wake of the financial meltdown, the site was put up for sale, but there were no takers, and several years later the building itself was demolished. Eventually, the land became the American Mobility Center, where a consortium of companies and government organizations have invested in a testing facility for self-driving cars and trucks.

The SPARK meeting is held in a conference room on the second floor of a garage that houses test vehicles. Moving across the room and greeting the other attendees before the meeting gets started, Rich eventually engages with Mike Martin, who runs one of Ann Arbor's premier realty companies. The conversation touches on some of the issues associated with a property Mike is trying to develop and Rich makes a mental note that Mike could be a valuable resource when Menlo needs to find a bigger home—something that might be necessary in the not too distant future.

8:30 AM
KEVIN F. & RICK COPPERSMITH
Developers

Back at Menlo's current site, Kevin F. and Rick Coppersmith meet at their shared workspace—a five-foot long table that's one of maybe 20 arranged in clusters of various sizes in the 24,000 sq. ft. basement of a parking garage in downtown Ann Arbor. Hanging from the ceiling are various cables and electrical cords that allow the tables and their hardware to be moved anywhere, which happens all the time as project needs change.

Aside from the bathrooms and the "baby room," the only enclosed spaces in the "factory" are three glass-walled conference rooms. There's a kitchen area, and adjacent to that, a carpeted area set aside for workshops and training sessions. Along one long wall is a cork "Work Authorization Board" covered with 5x8 inch "storycards" that lay out the discrete tasks associated with every current project. And there are posters everywhere. Robert Louis Stevenson: "Our business in this world is not to succeed, but to continue to fail in good spirits." Frank Zappa: "The computer can't tell you the emotional story. It can give you the exact mathematical design, but what's missing is the eyebrows." James Gregory Lord: "A vision is more than a concept or an idea. It is a force in peoples' hearts. Few forces are more powerful."

Grabbing coffee from the kitchen, Kevin and Rick get started on their day. It's still quiet in the room, but once everyone has settled in, the space will be filled with the hum of 50 or so people working together in close proximity. The sound is like a white noise where you can hear and be heard by those closest to you, while tuning out conversations that are farther away.

The two men, both thirty-somethings, present differently—Kevin, his hair long, wearing a gray sweatshirt—Rick, his hair cut short, wearing a plaid button-down. Kevin, only in his fourth month at Menlo, has an Associate's Degree in Computer Science from a local community college. Rick earned a BA in Chemistry and a Ph.D. in genetics at the University of Utah before switching gears and becoming a programmer. Different styles and backgrounds, but right now they sit side by side facing two monitors, one keyboard, and one mouse.

They start the day by discussing some notes they've received from the QA (Quality Advocate) folks working on their project, then turn to discuss a question from the pair of programmers working on the same project and sitting at the next table. McTavish, his tail wagging, stops by for a behind the ears scratch.

At 9:15, Kevin and Rick join Michelle, their project manager, and the other pair of programmers in one of the conference rooms for a brief, daily "stand up" call with the client and his programming team. Every Menlo project is assigned a code name, in this case "Quail." In Menlo's open space, with so many visitors coming in and out, the code names protect client confidentiality.

The Quail project involves building an application that will allow the client's B2B customers to use a more flexible set of on-line screens for posting their orders. The Daily Stand Up is a time to share updates, ask questions, and address concerns. Usually these calls run only about 15 minutes.

This one ends up running nearly an hour and a half, because it comes in the aftermath of what happened two days before when the client uploaded some data that the program was not intended or equipped to deal with. The program aborted, taking with it the client's online order-entry application. Not good!

As it turned out, Michelle, the Menlo project manager, happened to be visiting the client on that very day, and when the problem showed up, she immediately engaged the Quail team back at Menlo to help find a fix. She even arranged for pizza for the client's programmers, as they sweated out getting their order-entry app back up and running. Finally, after a couple of hours and with much back and forth between the two programming teams, the crisis was successfully resolved.

Today's Stand Up is devoted to analyzing the data entry problem and thinking through its implications. Rick keeps track of the conversation on a whiteboard, and when the call is finally over, his notes will be used to create a general game plan and split the necessary work among the two programming teams. In many other shops, a situation like this would have generated plenty of finger-pointing and defensiveness, but in this case, the client expresses nothing but appreciation for the way Menlo jumped in to help—including the pizza.

✦ ✦ ✦ ✦ ✦

As the members of the Menlo team head back to their workspaces, they pass some lockers that stand along one wall. From atop the lockers, two stuffed *Winnie the Pooh* characters—Eeyore and Tigger—peer down on them. But to Menlonians, these particular stuffed animals also represent two of their leaders. After all, James Goebel, Rich's co-founder and partner since Interface Systems days, admittedly shares certain characteristics with Eeyore, the somewhat dour and pessimistic (though entirely loveable) donkey who lives in the Hundred Acre Wood. And as for Tigger, he's a pretty good stand-in for Rich, with his ever (and sometimes perhaps overly) optimistic personality.

In a culture where one of the core operating principles is, "Expect leadership everywhere," every leader is fair game for a little good-natured teasing.

TWO HEADS ARE BETTER THAN ONE: PAIR PROGRAMMING

Two programmers, one computer, one screen, one keyboard, one mouse. The most prominent early advocate for pair programming was Kent Beck, author of *Extreme Programming*. Beck was one of the signers of the groundbreaking "Agile Manifesto," that urged a new set of principles and practices for software development. But Beck's passion for pair programming wasn't sufficient to convince the other signers to explicitly include it as an Agile practice. Some Agile organizations will have nothing to do with pair programming. Some will say, "We practice pair programming some of the time, like when someone first joins us, and we pair them up with someone more experienced for their first week." At Menlo, pair programming is the way it's done: two programmers, one computer, one screen, one keyboard, one mouse, every project, every day.

There are plenty of objections to pair programming. Programmers will say "I do my best work concentrating by myself on the task in front of me, not talking to someone else about it." Clients will argue, "You're charging me double for what could be done by one!"

But in fact, most studies of pair programming demonstrate that you don't get twice as much output from two coders working independently, as compared to two working in a pair. Pair programming often moves faster because when one of the programmers gets stuck, which inevitably happens, their partner will often come up with a suggestion. And a pair partner will often spot an error immediately, rather than the error only coming to light later after who knows how many lines of code have been written based on it. Pair programming also keeps a project from grinding to a halt if a programmer gets sick or goes on vacation.

And here's an absolutely crucial point. Pair programming works so well at Menlo because it is supported by the underlying axioms, values, and principles of the Menlo culture. Because that culture is truly focused on the common good, because it is truly committed to collaborative emergence—because the people Menlo hires want to work in a place that cares deeply about every individual but does so as a community—paired programming seems perfectly natural. And from a business sense, the proof is in the pudding: Menlo's clients love the products that paired programming delivers, which is why they keep coming back.

Pairing: Two developers, one computer, one mouse, one keyboard

8:30 AM
GUINEVERE PROJECT TEAM:
ANGELINA FAHS, GIOVANNI STURLA, HELEN HAGOS, KYLER WILKINS & LISA AND JOSIAH H.
Show & Tell

> *Deliver working software frequently, from a couple of weeks to a couple of months, with a preference to the shorter timescale.*
> -- A Principle behind the Agile Manifesto

Angelina Fahs, who lives just two blocks from the office, has the shortest commute of all the Menlonians. Just four months out of college, she's the youngest member of the Menlo family, and she's thrilled with her job as a developer. At the moment, she's also a little nervous, since she's going to lead today's "Show & Tell" for what has been code-named the "Guinevere" project.

In Agile software development (more about that later), projects are broken down into "sprints," typically a month or even a quarter long. At Menlo, each project is structured into what they call "iterations" of one or sometimes two weeks. Show & Tells, which take place at the end of each iteration, review what has been accomplished since the last meeting and offer an opportunity for the Menlo team and their counterparts at the client company to discuss problems they've run into and challenges they may run into going forward. Perhaps most important of all, this is where the client and the Menlo team reach agreement on, and commit to, what will be done in the next iteration.

In the Menlo scheme of things, Show & Tells exist to make sure that the client is getting what the client really wants and needs from the project. They're a big deal.

Making sure the client gets what they want and need, within the given time and budget constraints, is a responsibility embraced by the entire team. But the particular responsibility for tracking how the project is proceeding, especially with respect to time and budget, belongs to the Project Manager—in this case Lisa H. Lisa is a Menlo Mom, and as she and the team prepare for the Show & Tell, her six-month old baby Josiah snuggles quietly in a baby carrier against her chest.

In preparing for today's Show & Tell, Angelina has relied on all the other members of the Guinevere team, especially Helen, the most experienced developer on the team. Helen came to Ann Arbor from Ethiopia to study Computer Science at the University of Michigan, then stayed on to work at Menlo. Her guidance has given Angelina confidence that the Show & Tell will go well.

The Show & Tell is scheduled for 10:00. The client's project manager will be there in person, and the rest of his team will attend from their own offices via video link. At 8:30, the Menlo team is in one of the conference rooms, going through the storycards from this past iteration one more time. Each card delineates one task, and each is marked with a colored dot to indicate its status: *yellow* indicates that the team is working on the task, *orange* that the task had been completed by the developers, and *green* that the task had been checked by QA and could be officially considered "done!" A *red* dot means that some problem has occurred and work on the task has stopped.

As they review the work of the past two weeks, the team decides to take another, deeper look at storycard #266, dealing with a piece of code that has given them some problems. It worked fine the last time they ran it, but it won't hurt to check it again—and sure enough, this time the problem resurfaces. What's going on?

They run the code again and again, looking for the source of the problem, until finally Helen suggests, "Maybe the next integer will be what we need. Try running it one more time." And just like that, they've found the problem, and there's a ripple of light laughter as Giovanni quickly makes the necessary correction. There's real satisfaction—there's joy—at having solved the problem as a team.

That sense of working as a team extends beyond the walls of the Menlo factory to include the client's team as well. In this case, that tight working relationship is even more crucial, given the complexity of the Guinevere project, which involves three different client systems. To enable the two teams of developers to work together as effectively as possible, a phone link between the two teams is kept open all day long so that whenever a question or problem comes up, the line can simply be un-muted to allow for immediate collaboration.

In a sense, that open phone link mirrors the Menlo practice of having its project team members work in pairs and in close proximity to one another. It's a practical realization of *Enable collaborative emergence*, another core principle of a thriving organizational culture.

8:45 AM
JARON VOGEL & GRANT CARLISLE
Developers

Jaron Vogel heads over to the "Night Sky" project area, where two other developers working on the project, Al and Travis, have already picked a computer to work on for the day. Greetings are exchanged as Jaron plunks his backpack down on an adjoining table and takes off his coat, revealing a forest green flannel shirt.

Jaron is in his late twenties, with a neatly trimmed, dark red beard. He's definitely passionate about programming, and while he habitually wears a calm expression, a vein above his left eye seems to bulge when he's grappling with a complex problem. Jaron majored in Middle Eastern studies in college, including learning Arabic, a background that didn't come in especially handy in the landscaping jobs he held after graduation. But through all of that, he spent his free time learning how to program, until what started out as a hobby turned into a career goal, and finally (in 2016) to Menlo.

As Jaron boots up the computer, he spots Grant Carlisle, his pair programming partner for the week, coming across the room. Grant is in his mid-fifties, sporting a shaved head, clean-shaven face, and thin, wire-rimmed glasses. A software-veteran with decades of industry experience. Grant has seen, and definitely embraced the shift from traditional development practices to the more agile approaches employed at Menlo.

Grant smiles often, a smile that seems to completely take over his face. That smile breaks out as he greets his colleagues with a hearty "Good morning, Team Night Sky." Then he says, "I like your shirt" to Jaron, as he unzips his coat to reveal his own flannel shirt; his a black, white, and grey check.

Today, however, behind Grant's lightheartedness and bright smile, he is also considerably nervous. This is the end of the second week of his three-week interview process—an unusual hiring practice that doubles as thorough onboarding for those who ultimately are hired. At the end of the day, Jaron, Al, and Travis will give him feedback on how he's doing so far.

+ + + + +

Menlo's hiring process starts with a two-hour "Extreme Interviewing" event, which Rich Sheridan likens to a speed-dating session. Up to 30 applicants are brought in on the same day. They get a quick tour of the Factory and an overview of the Menlo business and the Menlo culture. And then, having been told that those who get through the day's event will be invited back for a second interview, they're put to work on three paper-and-pencil exercises, each 20 minutes long.

For each exercise, the applicants are paired with a different partner. The kicker is that they're told that their goal is to make sure that their *pair partner* looks good and gets invited back for the next interview!

As they work through each exercise, each pair is observed by one Menlonian. By the end of the process, each applicant has been observed by three Menlonians. As the pairs work, the Menlonians watch carefully to see how the applicants do on the so-called "kindergarten test." Do they share? Do they play well with others? Or do they hog the pencil? Boss their partner? Work on the problems all alone?

As counterintuitive as this process might seem—especially to the applicants who come in thinking about how to make *themselves* look good—it makes sense given the Menlo culture. That culture values things like collaboration and a commitment to the common good. It's not a culture built around individual superstars, but rather, around people who can self-responsibly work together as a team to create high-quality solutions that meet client needs.

Grant did well in the Extreme Interviewing event. He carried himself with confidence but was also quite humble and he exhibited the kindergarten skills that Menlo values. So, it was hardly surprising that Grant was brought back for a second interview, a one-day event in which this time he was paired with different Menlonians. As he worked with them on various project tasks, his partners looked to assess his technical as well his people skills. Again, Grant had done well, so he was then transitioned to the final stage of the hiring process—a three-week paid contract so that he could get a really good sense of what it would be like to work at Menlo, and Menlo could get a really good sense of what it would be like to work with Grant. Stage three is the make-or-break deep dive into an applicant's technical skills and cultural fit.

In the first week, Grant had been paired for a few days with Andy Russo, a senior developer, and then he had started working with Jaron on the Night Sky project. At the end of the week, Andy, Jaron, and a couple of other Menlonians who had observed Grant met with him to provide feedback on his technical skills, as well as his ability to integrate with the culture at Menlo. The feedback was positive: Andy gave Grant especially high praise for his project planning skills, commenting that Grant had shown a level of confidence in his ability to contribute to the team that can often take up to a year for new employees to cultivate. Jaron commented on Grant's comfort in switching from Andy's project to Night Sky, a valuable by-product of his deep industry experience, and an incredibly important skill at Menlo, where people often work on multiple projects during a single week.

VARIATIONS ON A THEME

Menlo's three-stage hiring process, including the final three-week engagement is used for all roles, but it is sometimes tailored to meet particular circumstances. For example, depending on the applicant, the three-week engagement may be shortened or lengthened. And rather than being contiguous, the three weeks may be spread out.

With all of this feedback about the strengths that Grant had shown in his first week, Andy and Jaron had also pointed out that Grant seemed relatively inexperienced when it came to "test driven development." With "TDD," developers write a test for a narrow piece of the source code, a test that the developers know the code as currently constructed can't pass. Then they write just enough new code to pass the test, repeating the process for each section of the program. TDD is, as Jaron commented to Grant, "an important part of how we [at Menlo] work." The message to Grant had been that in his second week, the team would look for evidence that while he might be inexperienced with test-driven development, he possessed the curiosity and the confidence to jump in the deep end and seek out opportunities to learn it.

8:53 AM
BRITTANY MORTON & RACHEL CLEVELAND
High-Tech Anthropologists

It's nearly 9:00, and the Factory is coming alive, with people going up to the "Work Authorization Board" to pull the storycards they will work on today.

STORYCARDS

Storycards are the building blocks of every project at Menlo. Each card spells out a specific task, written not in programmer language but in terms of a particular user need. At the top of each card is an estimate of how long it will take a pair of programmers to accomplish the task. Additional information, Dev Notes (Developer Notes), can be included, helping to ease the transition from one pair of programmers to the next on a project.

> Anatomy of a Storycard
> A Storycard:
> – Is an artifact for conversation...Is handwritten
> – Communicates business value
> – Describes functionality, not implementation
> – Can be estimated and tested

A poster on the wall at Menlo Innovations

The storycards are posted on the Work Authorization Board, which is marked off into projects and the days of the week. There are as many cards assigned to each day as it is feasible for the programming pair or pairs to complete. With each card marked with a colored dot indicating its status, anybody can walk up to the board and get a clear idea of the status of every current project. This is all part of Menlo's commitment to transparency, which in turn is critical to such core aspects of the culture as *connecting the dots* and *enabling collaborative emergence.*

Rachel and Brittany, two High-Tech Anthropologists (HTAs) are focused on an upcoming Show & Tell scheduled for 1:00. HTAs are Menlo's solution to the problem of software that comes with features end users don't want or can't use—however "cool" the developers may think they are. (*The Inmates Are Running the Asylum: Why High-Tech Products Drive Us Crazy and How to Restore the Sanity*, a much-referenced book in the Menlo library, is a harsh critique of how the software industry typically creates products that no one but other software engineers can love.)

HTAs, drawing on design thinking, go out to where the product will ultimately be used, to investigate the context, the practices, and the needs of end users. What they bring back feeds directly into the storycards that drive work on the project, and their continued input is critical to the final product's satisfying (delighting!) the client.

Rachel has only been with Menlo for seven weeks, and she's also new to the software industry, but she was recommended by her brother, who works at Menlo, and extolled her intelligence and her people skills. During the hiring process, and ever since, it's been clear that she's a fast learner and someone who will fit perfectly in the Menlo culture. As for Rachel herself, since coming on board, she's loved everything about the company and her HTA role.

Brittany is a variable-time worker whose hours depend on the business load. For Brittany, who has been with the company for two years, her flexible schedule is ideal, since it allows her to also tend to the jewelry business she owns and her avocation as a competitive swing dancer. As for her HTA role, she loves how it allows her to use her creativity in helping deliver great products that bring value and joy to the people who use them. If the definition of a thriving organization is one that enables its people to thrive, Brittany is a great example of how that works at Menlo.

Storycards on the Work Authorization Board

9:00 AM
MATT SCHOLAND & ART KLEIN
Quality Advocates

> *Working software is the primary measure of progress.*
> *-- A Principle behind the Agile Manifesto*

As the day begins, Matt Scholand and Art Klein start with the Quail Project. Unlike the developers, who typically switch pairs every week or two, Menlo QAs tend not to switch nearly as much, partly because of the nature of their work and partly because there are not many of them in the company. Matt and Art, in fact, work together some 80%-90% of the time.

Matt has been at Menlo for three years. He has a degree in criminal justice but found that was not for him. Before coming to Menlo, he worked as a bricklaying contractor, substitute teacher, baggage handler and at various other odd jobs. A friend at Menlo told him the company had a QA opening, he applied, and he's been here ever since. He has no formal programming experience, except for one course in college, but he's always been more facile with computers than the average user.

As a QA, Matt doesn't focus on the code, but rather on what the code is supposed to do and what it actually does. His job is to make sure the software delivers the outcomes the client wanted. He likens it to paying attention to what a piece of music sounds like, rather than analyzing the score.

Art has only been at Menlo for five months. He grew up in Ann Arbor, went to school at U of M, worked there and other places. He first learned about Menlo when Menlo allowed the Board of the

People's Food Co-op, of which Art was a member, to use the Factory for meetings in the evening. Art was intrigued by what he saw at the Factory, and his curiosity deepened over the years as he crossed paths with Menlo in other ways, including interactions between Menlo and the University of Michigan Business School, where he worked for three-and-a-half years in Academic Services. Eventually he applied for a QA job, and he's been really pleased with the work. In particular, he likes that he doesn't need to be "indispensable" as in some of his other jobs, where he was the only person who knew a certain process or content. In software development jargon, at Menlo he isn't a "Tower of Knowledge."

A "Tower of Knowledge" is an often-used term in the software industry to describe the one individual who is the *only* person in an organization with a deep knowledge of a particular code or process. They may very well have written the code or installed the process in the first place. And, in any case, over time they've upgraded the code or tweaked the process until they've become the only person capable of making new upgrades or fixing things if something goes completely haywire. Being in that position undoubtedly has a certain appeal—after all, everyone wants to feel needed—but being in that position also creates its own pressure. What happens if you're out sick or on vacation when your unique knowledge is suddenly needed? How will your colleagues deal with the problem? How will customers be affected? Can you really be away from the office?

At Menlo, where people routinely work on projects that require programming languages and tools that are new to them, and which they learn on the fly from the partners with whom they're paired, nobody is or has to be under that kind of pressure, and that's just fine with Art.

Anyway, today the first thing on the agenda for Matt and Art is to test a feature of the Quail software that matches payment and invoices. This is currently being done on the client's outdated legacy system, which needs to be retired. Matt and Art have constructed a test to troubleshoot this process, but the QA computer they're using has such a large database that it will actually get in the way of the test. To save time, they decide it would be best to switch to a Development (Dev) computer, which has much less data. So they go over to a nearby table where a pair of programmers are working, talk the idea through, and finally move over to an open Dev "box" to run their test. (Note that computers at Menlo are set up for specific roles—HTA, QA, Dev, etc.— and the computers are assigned to projects, not people. No one has a personal Menlo computer or laptop—except Rich and James have laptops they use for talks and presentations.)

As they run the test, Matt and Art go back and forth, constantly asking for each other's take on what they're seeing, trading turns at the keyboard, and sometimes lapsing into an intense silence as they both think about what they should do next.

To an observer, the process seems calm yet highly focused, effective without any obvious signs of pressure or frustration. This same demeanor will in fact characterize their work throughout the day as they move on to other tasks.

When Matt and Art finish running the Quail test, they put a green dot on the storycard for that task, showing that it is QA approved. Then it's on to the Cherubim project. At the Work Authorization Board, they check the Cherubim "swimlane" and note the card that shows that the software needs to be tested to make sure it works in Internet Explorer (IE). They prioritize this particular card because of the 1:00 pm Show & Tell with the client, and because they can tell from the Board that the software has previously failed this particular test and been re-worked by the developers. Because they can see what's already been done, and because they understand the priority of client-focus, they can make this decision on what to work on without asking for the PM's approval.

MAKE DECISIONS AT THE LOWEST LEVEL

In high-performing organizational cultures it is increasingly the case that more and more consequential decisions are being made by the "doers" who experience the issues most directly.

They go back to their table, run the test again, and this time the software performs perfectly. It too gets a celebratory green dot.

✦ ✦ ✦ ✦ ✦

In most companies, "QA" stands for Quality Assurance. So where did Menlo's "Quality Advocate" come from? Neither Matt nor Art knows for sure. "I think the point, says Art, "is that 'Quality Advocate' frames the quality work in a different way. We're working *with* the developers, not against them. We're not adversaries." To which Matt adds, "Quality Control is about the end-product and Quality Assurance is about the process. With Advocacy, we wrap both outcome and process into one."

Art calls out to Marilyn, thinking that with her long experience at Menlo, she might know. "Marilyn, do you know where the term Quality Advocate came from?" Marilyn says that she doesn't know for sure, but "But I'm sure it was to indicate that these Menlonians were looking out for the client."

James Goebel, who is sitting nearby, jumps in: "Yes, it was early on and it was to differentiate what Menlo does compared to other organizations. If they were called Quality Assurance, then everybody would already know that term and how it fit into their world view and their other experiences. They would say to themselves, 'got those' and wouldn't realize that what QAs do at Menlo is different from what QAs do in almost all other companies. The same thing was what led to the name High-Tech Anthropologist. The term resonated and had enough difference to open up a conversation about what it was. People would hear the term and say, 'Oh yeah, yeah… What is that?'

"Also, there's this tendency for people to think that the problems QA folks identify are somehow their fault. You know…THEY found this problem. We didn't want that adversarial orientation. That's part of the reason why Menlo doesn't have a separate QA pod, but instead has the QAs sitting next to the developers and others working on a project."

9:00 AM
JARON VOGEL & GRANT CARLISLE
Developers

Grant and Jaron have pulled the storycard for today's first task. The Night Sky application has been in development for several months, but it's still not at a stage where it can display critical data in any useful fashion. Basically, the formatting is a mess, and no user would actually be able to glean anything from the way the data is currently presented. Today, Grant and Jaron are going to fix that problem.

Nearly all Menlo projects include a front-end consultation period during which a pair of HTAs will spend anywhere from a couple of weeks to a couple of months observing the people who will eventually be using the new software. The HTAs produce detailed user profiles to figure out the features and functionality that are really needed. Then they build a set of mockups, dummy images of what the software's screens might actually look like, highlighting the various features and giving the programmers a better sense of the desired user experience.

Depending on the project, this HTA consultation period can be expensive, and some clients can't or just don't want to make that investment, and that's that. Generally, Menlo won't accept the project. Sometimes, although not often, Menlo will take on the project anyway, foregoing the HTA involvement. Night Sky is such a project, and that's creating some problems, in particular, that the mockups the client has provided are not matching up with feedback from the client's project manager (PM).

The client PM, for example, has told the Menlo team that the displays on every screen, throughout the application, should use lines—the same kind of lines—to separate each row of data. The mockups the team received from the client, however, show lines in some parts of the application but not in all. The mockup for the screen Jaron and Grant are working on right now has no lines, so what should they do?

Grant says they should go with the mockup and leave out the lines. Jaron responds, "Yeah, I get that the client didn't put lines in the mockup, but I think we should put them in anyway. Let them tell us explicitly that they don't want them."

"But if they wanted lines, wouldn't they have put them in the mockup?" asks Grant.

"Not necessarily. Clients are not always as careful as our HTAs to really dig into what the end users want and need. That's why it can be a problem when they don't use the HTAs. If we listen to the PM and put in the lines, the client will either be fine with that or they'll push back, and that would be a good conversation to have."

"Got it," says Grant. "Works for me." Thirty minutes later, they've put lines into the display and they're ready to move on.

AGILE...AND SO MUCH MORE

In 2001, right at the time Rich Sheridan and James Goebel were getting Menlo off the ground, 17 thought leaders in software development met at the Snowbird ski resort in Utah and penned their famous *Manifesto for Agile Software Development.* The Manifesto declared the importance of "individuals and interactions over processes and tools;" "working product over comprehensive documentation;" "customer collaboration over contract negotiation;" and "responding to change over following a plan."

The Manifesto also laid out some working principles, many of which are still mainstays of Agile development. Quick iterations of working software (known as "sprints") with feedback from the customer. Daily Stand Up meetings to ensure that everyone on a project team knows what's going on. Businesspeople and developers working together on projects. Self-organizing teams. Lots of face-to-face interaction. Time for reflection to learn from experience. And so on.

This all sounds very much like Menlo, which raises the question: Is Menlo just a good practitioner of Agile practices?

In "Menlo vs Agile," a blog on the company website, Corissa Niemann notes that, "From Lean to Agile, we are known for borrowing the tools and practices that we find useful and making them our own." She points out that "Many of the differences 'between Agile and what goes on at Menlo lie almost solely in naming conventions." Agile talks about "user stories;" Menlo says "storycards." Agile talks about "sprints," Menlo about "iterations." Agile's "sprint planning meetings" and "scrum boards," are "nearly identical to Menlo's "Kick-Off meetings" and "Work Authorization Board" respectively.

But there are differences. "Daily Stand Ups" in most Agile organizations are team-based, while at Menlo they involve the entire organization. So instead of just 5-10 people, Daily Stand Ups at Menlo involve maybe 50 people, with even visitors frequently joining in. Daily Stand Ups at Menlo aren't just about getting an overview of what's going on with your own team, it's about getting an overview of what's going on in the entire company.

And where sprints in most Agile organizations typically last a month or a quarter, Menlo iterations are much shorter, just a week or two long. Where Agile organizations typically assign an internal businessperson to bring the client's perspective to project debates, Menlo brings in the actual client, and maybe the entire client team, for the Show & Tell after every iteration. And while it's common in Agile organizations for a team to participate in hiring decisions, at Menlo many more Menlonians will typically be involved.

But there is something even deeper about the differences of Agile at Menlo than at many other companies. Let's be honest. "Agile" is currently a management buzzword. Who wants to be known as sluggish, sclerotic, or bureaucratic? Unfortunately, too many companies tack on one or two "Agile practices" (like a Daily Stand Up) and think—job done! —they're doing Agile! But that doesn't necessarily make them customer-focused or nimble or adaptive or flexible—all the things that following Agile methods is supposed to make happen.

At Menlo, Agile definitely isn't a buzzword. It isn't even just a handful of practices. Instead, Menlo embraces a disciplined, deeply principled, and profoundly pervasive approach that takes Agile methods to new heights.

So, does Menlo do Agile?

The question makes Rich Sheridan bristle a bit because it connotes another "flavor of the month" management trick and not a complete cultural mindset. Rich says, "I never liked "doing" Agile. He makes the important distinction between "doing Agile (with a capital "A" to denote the method) to "'being agile (with a small "a" to denote adaptive flexibility.'" He adds, "I like to think Menlo delivers value using the power of Agile… and so much more!"

9:00 AM
BRITTANY MORTON & RACHEL CLEVELAND
High-Tech Anthropologists

Behind all decisions related to client work stands Menlo's mission statement. It is not only known by all Menlonians, it is lived constantly. It informs choices and decisions. It is a guiding light. It speaks to their work, their clients, the end users of their product, and the Menlonians themselves. Here it is:

Our Mission

"End suffering in the world as it relates to technology"

We do this by focusing on three different stakeholder groups in our industry:

Software project sponsors who traditionally have had little hope of steering a project to a successful conclusion before money and executive patience is exhausted.

End users of the software who, far too often have no voice at all in the design yet must live every day with decisions of people they have never met.

The software teams themselves, who typically labor under years of overtime, missed vacations, and unrealistic expectations only to have their projects canceled.

Return joy to technology.

✦ ✦ ✦ ✦ ✦

As they prep for today's Cherubim Show & Tell, Brittany and Rachel have quickly gone through their to-do list. New to Menlo, Rachel has only sat in on a few Show & Tells, but she's going to play a more active role in this one, so she's excited and a little nervous, and glad that Brittany will be right there with her.

After reviewing their list, they head to the Work Authorization Board for another look through the storycards for this iteration of the project. After discussing each of the cards, and satisfying themselves that the team has done the work it committed to at the last Show & Tell, it's back to their table to work on the PowerPoint they'll use at the meeting.

Rachel and Brittany are not the only HTAs who have worked on Cherubim. In fact, just yesterday Rachel was paired with Mollie Callahan, who has been involved in the project from the beginning. As they work on the PowerPoint, Rachel and Brittany occasionally turn to Mollie, sitting at a nearby table, for her input.

As important as Show & Tell is, as the prep work goes on through the morning, Rachel and Brittany seem calm. They give off the sense that "We've got this."

+ + + + + +

HTAs typically write the first set of storycards for a project, based on their deep dive into who will actually be using the new software and what they need it to do. Menlo's stated (rather grandly stated) mission is to "End human suffering in the world as it relates to technology." With that in mind, the HTA's priority is always how to maximize the end user's satisfaction and minimize their suffering when using the product Menlo delivers. Critical to that task is identifying who will really be using the software.

One of the problems with many development projects is that the client—typically a senior manager or executive who very often is *not* the primary end user—asks for too many features, thinking of all the people who might possibly use the software. All too often this leads to scope creep, to products that take too long and cost too much to deliver (and sometimes never actually get delivered), products that are too complicated for the main purpose they are actually intended to serve.

By helping the client identify the main user, the "primary persona," the HTAs help focus the work. It is certainly possible that the client will decide that the software should also work for "secondary" or even "tertiary persona," and in such a case the HTAs will research those folks as well. But being clear about whose needs are primary and whose are secondary or even only "nice to have" brings real clarity to discussions about what can be accomplished in what timeframe and at what cost.

✦ ✦ ✦ ✦ ✦

At 10:00, as Brittany and Rachel are working away on the Cherubim PowerPoint, someone yells, "Hey HTAs!" and then shares something he has discovered that he thinks would be useful for the HTAs to know. This calling out—"high speed voice technology" as Menlonians refer to it—happens fairly often at Menlo, which tends to rely very little on internal texting or e-mailing. Sometimes an individual Menlonian is called out, sometimes the call targets a project team. Sometimes someone will even shout "Hey Menlo!" to reach everyone.

To outsiders, this all sounds as it might be pretty distracting, but Menlonians have learned to largely tune out the call-outs that are not aimed their way, and they seem to regard the practice as something that's not only fun, something that makes Menlo a special place, but also as something that contributes to the close connection that exists among them all, regardless of their role or pay grade. That's organizational culture at work.

9:00 AM
MOLLIE CALLAHAN
High-Tech Anthropologist

Mollie Callahan, a High-Tech Anthropologist, has a warm, welcoming way about her that makes other people instantly feel comfortable. She obviously cares about people and about the work she does. James Goebel sometimes quips, "There are three extroverts working at Menlo." Mollie is one of them.

Mollie is also a "real" anthropologist, with a Ph.D. in Anthropology, and a two-year post doc position teaching at Oberlin College and conducting field research in South America. Although she was gratified by her relationships with her students and with the people she studied, she was much less inspired by the idea of "publish or perish." So, she moved on to Teach for America, teaching middle school science in inner city Dallas, before transitioning into the private sector.

Three years ago, she came to Menlo. She believed then, and still believes, in the idea of ending human suffering as it relates to technology. That's the kind of tangible impact on people's lives that she's always wanted to have.

Mollie's technical training in Anthropology isn't all that useful in her HTA role. She even points out that when she first started at Menlo, her tendency to use the language of a professional anthropologist was a hindrance, rather than a help. On the other hand, a good field anthropologist has keen observational skills and a strong sense of empathy, and those attributes are definitely valuable in her work. HTAs have to closely observe end users, looking to understand what they do, how they do it, what problems they run into, and so on. Their focus is always on real, specific end users. As one HTA put it, "If we're trying to solve a problem for everyone, we're solving a problem for no one."

Right now, the big space around Mollie is starting to buzz, even though only about half of the Menlonians are at their tables working. With the high value it places on trustworthiness and self-responsibility, Menlo's culture means that it's taken for granted that everyone will put in the time necessary to get the work done, and typically, pair partners coordinate their arrival time to make sure neither partner has to wait for the other to get started. As more and more people come through the front door, no one seems especially rushed, but they waste little time getting down to business.

> **FOCUS ON LONG TERM SUCCESS**
> Thriving organizations understand that they need to take care of today and tomorrow, but they also never lose sight of the need to prepare for a future they can't fully anticipate.

Mollie is actually working by herself today. She's working on Project Liberty, one of four internal projects—Liberty, Freedom, Justice and Independence—focused on making Menlo sustainable, financially and otherwise, well into the future. Freedom, Justice, and Independence are all about making sure Menlo is attracting future talent, providing comparable salary and benefits to all Menlonians, and charging clients appropriately. Project Liberty is focused on shifting Menlo's sales from reactive ("the leads will come to us") to proactive ("we're going to the leads").

✦ ✦ ✦ ✦ ✦

Project Liberty was launched in 2017, following what Rich refers to as "the Sum of All Fears meeting." Menlo practices Open Book management, which involves weekly meetings at which the entire organization looks at all the relevant financial data that indicates how the company is doing. One of the key metrics tracked at these meetings is the projected number of weeks the company can make future payroll, given cash on hand plus the business in the pipeline. Over the years, that metric had consistently looked good, growing to the point where the company's future seemed secure. The question had in fact become how to balance growing that financial cushion against the need to hire more programmers. At what point did the desire for an extended cushion mean Menlo would have to tell potential clients that it would be some time before they could start work on their project? And at what point would that send a potential client looking elsewhere?

But in 2017, Menlonians watched as, week by week, this crucial metric declined. There wasn't enough business coming in to keep paying everyone very far into the future. And as what had been a comfortable cushion grew less and less comfortable, people grew more and more fearful. Jobs, maybe even the company, were on the line. At an Open Book meeting in October—the Sum of All Fears meeting—the fear seemed to reach a crescendo.

Fear has no place at Menlo. Right from Day One, Rich and James have worked hard to keep it out. In the Menlo culture, there's no playing favorites, no intimidation, no hiding of information. But here was a situation where the open sharing of important information was actually creating fear. As Mollie describes the situation, "It was a rough time, and then it started to get scary rough."

In response, the team decided that there was a need to reevaluate and reconsider Menlo's approach to sales. When Menlo was first founded, one of the then four co-founders had gone ahead and hired a salesperson. He had made that decision without the consensus of the other founders, which was a problem in itself, but the problem was exacerbated by the fact that a traditional commission-based sales model didn't turn out to be a good cultural fit at Menlo. In trying to maximize his own income, the sales guy routinely closed on projects regardless of Menlo's ability to deliver at the necessary level of quality or profitability. So much for creating joy and eliminating suffering associated with technology.

That was the end of Menlo's experiment with a traditional sales model. And over time, as people heard about Menlo through the business grapevine, business had indeed come in over the transom. It might be a long time, even years, between the first time a company heard about Menlo and the time they actually initiated a project, but for a long time, the pipeline had stayed comfortably full. But in 2017, the pipeline seemed to be drying up. Was it time to re-try the idea of a dedicated sales force?

As the discussion played out, Rich reflected on an incident that had somewhat haunted him. He had gone to lunch one day with the CIO of a large company, someone he had known for a long time, and with whom he had had fairly frequent contact—but as yet, no contracts. So that day, Rich had said to him, "You know, we've had you and any number of people from your teams on tours at Menlo. You clearly seem intrigued by what we do. But it has never turned into a project. Why is that?" In reply, the CIO had said, "Rich, you never followed up with any proposals."

For Rich, that was *mea culpa*. He had sent bland follow-up emails after various contacts, but he had never specifically asked the CIO if there was a project Menlo could work on for him. Clearly, it was time for a different approach.

As it had turned out, the approach was not to bring in dedicated salespeople. Instead, half a dozen Menlonians went off for sales training. Jane went. So did Mollie and Andrew and some others. What they came to realize was that their knowledge of the Menlo approach to software development plus the personal skill sets of the HTAs and Project Managers—combined with some basic sales disciplines—actually created the basis for an organic Menlonian "sales function" that was mission and culturally consistent.

Project Liberty was created to follow up on that insight. The idea was to focus on intentional follow-up with people who came to Menlo for tours and workshops, people who came to events at which Rich spoke. The idea was to be more proactive, to follow up more

INVEST IN FUTURE (ADAPTIVE) CAPACITY

While almost every organization invests for growth, often these investments focus on extending what the organization already does well. Thriving organizations do that too, but they also invest to learn about things that may or may not be useful in the future.

systematically, and equally important, to involve people other than Rich in pursuing external contacts. Instead of Rich being a Tower of Knowledge when it came to sales, which is what had inadvertently happened, Project Liberty was an investment in building a disciplined, systematic process for capturing, documenting, and distributing knowledge related to potential clients.

✦ ✦ ✦ ✦ ✦

As she settles in for the day, Mollie checks in with a nearby colleague. She talks about the reading she went to last night at the Literati Bookstore to hear Annie Lamott. She explains why she loved

Lamott's latest book, and says, "I think you'd really like the chapter called 'Shitty First Drafts.' It's all about how the first time you do something, it never turns out perfect. You have to go through a shitty draft first."

"Thanks," her colleague responds, "I'll get a copy." Then they both turn back to their work.

Rich sometimes talks about how one of the bosses he worked for during his "corporate life" had told him, "When you see two people talking, that means they're not working. So what you do is just go up and stand next to them and stare. Soon they'll get back to their desks. Works every time." Rich admits, candidly and sadly, that he learned that lesson, and developed some finely honed intimidation skills back in the old days. Now he sees things very differently. Those casual human interactions during the workday contribute to a healthy environment. They're a natural, and positive, aspect of a culture that treats everyone with respect and places a premium on trustworthiness and self-responsibility. They don't detract from performance, they enhance it.

In a way, Project Liberty is about extending Menlo's internal environment to the outside world. It's about developing an approach to sales that's based on really understanding and meeting the needs of potential clients—about building trusting relationships with them. It's about articulating in a convincing but not pushy way what Menlo might be able to do for them.

Given Mollie's outgoing personality, along with the professional empathy she practices as an HTA, she's a natural fit for the Liberty team, which includes people from every area—HTAs, developers, QAs, and the Experience team. The project runs like any external project, including its own weekly—and widely attended—Show & Tells.

Mollie's first priority is to follow up on a trade show she and David Crenshaw had recently attended. She looks around and finally moves to share a table with Asha, another HTA, and adjacent to a table where Brittany and Rachel are working on Project Cherubim. Mollie was involved with Cherubim in its early days and figures she might be able to lend them some support as they get ready for their Show & Tell. She wouldn't intrude, but she'd be there to help if needed.

The urge for coffee makes itself felt, so she heads back to the kitchen area, pours the last of the coffee from the pot, and makes a fresh pot. At Menlo, keeping coffee in the pot is everyone's job. If you take the last cup, you make the next pot. Simple. Noticing that the coffee supply is running low, she makes a mental note to let Anna know. While Anna is the de facto office manager, she's not expected to be a Tower of Knowledge either.

Back at her table, Mollie filters through the emails that came in after she left for the day yesterday. Menlo prides itself on respecting the boundaries between work and home life, so neither Mollie nor most any other Menlonian checks their work-related email at night. One of Mollie's emails is related to Project Freedom, and as she's reading through it, Anna walks by and lets her know that pizza is being brought in for lunch. And then their conversation takes a different turn.

9:30 AM
JANE COLLINS
Project Manager

Jane Collins is running late. It's 9:30 when she walks into the Factory, cup of coffee in hand, and sits down at a table near those occupied by Rich and Anna. She looks comfortable, wearing a black sweater and blue jeans with her blond hair in a ponytail, but Anna asks: "Everything all right?"

"I'm not feeling well," Jane replies. "Probably going to leave early today."

"That is probably my least favorite thing about the open space," Anna says, jokingly. "We work together so much that once someone is sick, everyone is sick. That's why I keep this giant bottle of hand sanitizer at my desk. It's the only way you can stay healthy around here."

Jane laughs and begins checking her email, some of which pertain to her role as a Project Manager, others to her additional role as Menlo's Controller. She started at Menlo as a PM, moved into the Controller job in 2015 when the CFO retired, and has served in both roles since 2018.

As a PM, Jane is the primary point of contact with the client as parts of Menlo's developers and HTAs change through the duration of the project. She tracks progress on the project and works with other PMs to make sure that each project is staffed with enough pairs to stay on schedule.

As Controller, she handles an array of financial and administrative responsibilities—paying bills, tracking payments, processing payroll, as well as various other *ad hoc* tasks. Recently, for example, she helped

Rich and James reach the decision that in the coming year, Menlo will pay 100% of the premiums on the HMO health insurance option. That decision was just announced yesterday.

While Jane enjoys both of her roles, she acknowledges the challenges that come with juggling both. She's had to engage in some very un-Menlo practices like staying late or working on weekends. She'd like to hire someone to help with her workload, but she appreciates that there's a more general pressure to increase the workforce to keep up with the project load. She also appreciates that Menlo is cautious about growing too fast, in part because of a concern that bringing in new people will somehow change the culture.

Like many other Menlonians, Jane's path to software and to Menlo was anything but straight. She went to graduate school in Boston where she got a Ph.D. in Astronomy. Her boyfriend at the time graduated a year before she did and took a position at the University of Michigan. When Jane graduated, in 2006, she accepted a post-doc at the University of Michigan and joined him in Ann Arbor.

After a year and a half, Jane decided not to renew her post-doc. Her research was satisfying, but it was also high-pressure, required long hours, and people generally worked on their own with minimal collaboration. All in all, "it was not a very joyful position," Jane says. By then it was 2008, not a great time to go job hunting, but Jane left the university anyway. She ultimately used the time off to plan her weddings. Her fiancé was from Bulgaria, so they would have one wedding in the U.S. and one in Bulgaria.

Sometime later, Jane met Megan, one of Rich's daughters. Megan was leaving a PM position at Menlo and asked if Jane would be interested in taking over her role. Jane said yes and that's how she got to Menlo. Serendipitously, her astronomy background even proved to be an asset on one of her first projects, which involved algorithms she had used in her dissertation!

When she talks about her eight years at Menlo, Jane talks about how success at Menlo is defined as collective success, not individual achievement, titles, and so forth. She also marvels that she's never bored at Menlo. There's always something new for her to work on and ultimately, new ways for her to help the people she works with. Of the future, she says, "I think I'll still be working here in ten years, even if I don't know exactly what I will be doing."

✦ ✦ ✦ ✦ ✦

Today Jane needs to work on payroll. The payroll data has to be submitted to the payroll processing company before noon on Monday. Menlonians keep track of their time in 15-minute intervals, and they do it on paper, which seems a little strange for a software company! At the end of the week, Jane compiles the data and Carol Sheridan—the "Factory Manager"—enters it first into QuickBooks, and then into an Excel spreadsheet for the payroll company. There's something about the physical act of writing down what they've been working on and for how long that seems to make Menlonians more conscious of what they're about.

The accuracy of those time logs is crucial. Even though Menlo provides estimates of how long each project task (and thus the entire project) will take, and therefore how much it is likely to cost, clients are actually billed for the exact hours worked. Clients agree to pay for actual time worked, regardless of the initial projected cost.

When he talks about Menlo's billing method, Rich tells the story of how he responded when one potential client said, "You mean if it takes longer than you estimated, I have to pay for it?" Rich fired back, "What do you think we'd do if we always had to be sure our estimates covered our costs?" After a moment, the answer came back, "You'd probably always estimate high, I guess." "Exactly," chimed in James, the COO, "and if the work took less time than we predicted, would we give you your money back?"

Of course, the reason clients are okay with this arrangement is that they trust that the hours on the invoice are in fact the hours spent on their project. As with so much else at Menlo, trust is crucial, both internally and externally.

> **TRUSTWORTHINESS**
>
> Trust isn't an input, it's an outcome. You get trust by being trustworthy. Being trustworthy generally means that you operate with good intentions, that you're competent at what you do, and that you act ethically.

As for the Menlonians, they are all—except Rich and James—paid by the hour, and they all qualify for overtime. But before someone stays late working on a particularly demanding project, the client has to agree to pay for the overtime. And generally, Menlo's ethos of real "work-life balance" discourages late-night work, except to deal with an issue that the Menlo team and the client perceive as extremely pressing.

It's not uncommon for errors to occur as the payroll data is entered into QuickBooks and transferred to Excel, so Jane is very careful to ensure the accuracy of the final output. It matters to the clients, of course, but it also matters to her fellow Menlonians, who want and deserve to be paid accurately. The fact that Jane knows everyone on the Menlo payroll helps her to keep everything in good order.

Several computers in the office have QuickBooks installed on them. Jane uses the admin account on QuickBooks for her day-to-day work. When she's sick or out of the office for some other reason, Rick can use the admin account and fill in for her. Anna enters company credit card receipts in QuickBooks. Carol generally enters the timesheets. All of this helps ensure that as much as Jane has primary responsibility for this work, she is not a Tower of Knowledge.

Menlo's QuickBooks system is open to all Menlonians (although not everyone has editing access). This information feeds into the weekly Open Book meeting, so that everyone knows how the company is doing from a financial standpoint. This is not only an outgrowth of Menlo's commitment to the common good, trustworthiness, connecting the dots, and other axioms, values, and principles of a thriving organization It's also a way to encourage every Menlonian to "think like a businessperson."

9:30 AM
RICH SHERIDAN
CEO and Chief Storyteller

Now that everyone has had a chance to grab coffee and maybe a pastry, Paul Krukto, the President and CEO of SPARK, launches into a PowerPoint presentation on what has been accomplished since their last meeting.

The A2Tech360 meeting was "a great success." (Ann Arbor is often abbreviated as A2.) There's been significant positive momentum on the American Mobility Center. Cisco has just acquired DUO—an A2 computer security start up for $2.35 billion, the second largest acquisition Cisco has ever made. And KLA Tencor—a Silicon Valley company supplying process control and yield management systems for chip manufacturing—has announced that they are going to build an R&D facility in Ann Arbor that will bring in about 500 jobs.

Next Paul shows a slide of a recent cover of the *Economist* with the headline "Peak Valley" and a story about why startups are moving out of Silicon Valley. "Everything there is too expensive," Paul says. "Everybody knows that. What most people don't know is where the highest concentration of IT talent outside of Silicon Valley is located. You might guess Boston-Cambridge, or Austin, Texas. But you'd be wrong. Southeast Michigan has the second highest concentration of IT talent in the US. And you know why that's important? Sixty-six percent of all companies identify finding, hiring, and retaining quality talent as their biggest concern."

Next, Paul announces that Ann Arbor is one of only three cities to recently be awarded a second 15-year tranche of money from the State of Michigan for supporting startup activity, as part of a program known as SmartZones. Rich cracks, "Does that make us a SmartestZone?" and everyone laughs.

Paul then segues to the day's main topic. The idea behind the American Mobility Center is to build a "real world" test environment for autonomous vehicles, a set of roadways that incorporate all of the typical features that might challenge a self-driving car. Those features include sections that are like high speed interstate lanes, and an intersection of three lanes in each direction, the kind of location that produces the most traffic accidents in the US, according to the National Highway Safety Traffic Administration (NHSTA). There are sections with cargo trailers stacked on top of one another to simulate driving in dense urban areas, and a stretch of dirt roads to simulate rural or off-road experience. (The rural experience currently comes with a family of three deer that the Mobility Center is planning on catching and releasing in some safer location elsewhere in the state.) There are a couple of three-tiered highway sections (underpass, main roadway, and overpass) to challenge vehicle location sensors, and a 750-ft curved tunnel which incapacitates the car's GPS signals.

As Paul describes the tunnel, Rich calls out, "Did they ask if you could have it be snowing as the cars exit the tunnel?" Again laughter, but Paul replies, "Actually, the fact that we get snow does make us a better location than some of the California sites where it's all sunshine. And we have a special rain truck that can blow anything from a light mist to a heavy downpour onto the test vehicles."

Paul closes with a slide showing all the companies that have been involved so far in the consortium— companies like AT&T, Siemens, and Microsoft as well as the expected automobile manufacturers like Ford, Toyota, and Hyundai. While Rich thinks the advocates of self-driving cars are underestimating how big a challenge it will be to make them fully autonomous, he also appreciates that creating the software to meet that challenge will represent an opportunity for companies like Menlo. In that sense, his involvement with SPARK is not only about promoting the common good, it's also an example of the cultural principle of cultivating a savvy sensitivity to the outside world.

✦ ✦ ✦ ✦ ✦

The group has taken a bus tour of the Mobility Center. Now they've come to a stop near two test vehicles, a black Lexus and a white Ford Fusion. Rich was sitting near the front of the bus, so he's one of the first in line for a test drive.

He climbs into the back seat of the Ford Fusion. There are two technicians in the front seat, monitoring the car's progress, but the car is definitely driving itself. In the back seat is a display that shows the car's path through a set of obstacles (strategically placed cones). The car navigates through the cones, accelerates down straight stretches of roadway, and comes to stops—appropriately—at stop signs. Along one straight stretch, a broken-down car on the shoulder encroaches several feet into the lane. On the display, the car appears as a yellow icon, and Rich can see how the test vehicle nicely swerves around the obstacle and then back into its own lane.

With his test drive concluded, Rich has time to chat with technicians about some of the navigation and tracking systems before it's time to head back to the parking lot and his own car. So far, it's been an interesting morning.

9:30 AM
MOLLIE CALLAHAN
High-Tech Anthropologist

When she glances down at Mollie's computer screen, Anna sees that she's reading a Project Freedom email about a potential intern. Anna, who is also working on Freedom, says, "What do you think?"

"I'm not sure he's be a good fit for us."

"Yeah, Michelle and I have the same feeling. He's really green."

Project Freedom is another of the projects that came out of the Sum of All Fears Meeting. The project is focused on recruiting and retaining highly skilled new people, in keeping with the core cultural axiom of *Focus on Long-Term Success* and the related principle of *Invest in Future (Adaptive) Capacity*. The announcement yesterday that going forward, Menlo will pay 100% of the premiums on the HMO health insurance option is one outgrowth of Freedom. By making Menlo more competitive with other similarly sized companies in the area, it should help with both recruitment and retention.

The person mentioned in the email was put forward to Menlo by an organization that helps individuals acquire needed job experience by arranging, and subsidizing, apprenticeships in local companies. With its commitment to the local community, Menlo is trying to work with the organization, but Project Freedom is aimed at recruiting high-level talent, and unfortunately, this individual doesn't meet that standard, at least not now.

✦ ✦ ✦ ✦ ✦

As Anna walks away, Mollie turns back to her follow up on the trade show. The Project Liberty team is looking at trade shows as a possible sales tool, and Mollie has been tasked with starting to collect data on the subject. Menlonians love to run experiments, and this trade show foray is just that, another experiment.

The show the team had picked for this experiment was focused on automotive manufacturing, an industry in which Menlo already has a toehold. As an HTA, Mollie has worked on several projects with automotive companies. Her goal for the trade show had been to track how many leads and potentially valuable relationships Menlo might pick up by having a booth, and to develop some ideas about how best to present Menlo's message to stimulate those leads.

Mollie starts by sending photos from the show to the other Liberty team members. Then she digs around in the oversized tote bag she carries almost everywhere and pulls out several plastic baggies in which she had dropped the notes she jotted down at the show.

To better collect and track contact information at the show, Liberty invested in an app that allowed Molly and David, a co-leader on the project, to simply scan a barcode on an attendee's name badge, and then just click to input data such as the strength and interest of the lead. Today Mollie plans to download the information on the app, along with the information on her notes, into HubSpot, Menlo's CRM (customer relationship management) system. She can then go through the data to identify the hottest leads, discuss everything with the other Liberty team members, and develop a follow-up plan.

But then, unfortunately, when Mollie tries to log into the contact app, she gets the dreaded "Username or password is incorrect" message.

In case she mis-typed the log-in information the first time, she tries again, more slowly. Nope. Same error message.

After trying a few random variants, and getting the same error message every time, she finally locates the system manufacturer's customer service number and calls the help desk. Pushed into voicemail, she leaves a message.

✦ ✦ ✦ ✦ ✦

"Mollie, got a minute?" It's Brittany, from the adjacent table where she and Rachel are getting ready for the Cherubim Show & Tell. "We've got a question about one of these storycards."

Mollie spins in her chair. "What's the issue?" she asks. "We think there was an omission on the card," Rachel replies, "and we're not sure how we should account for the work."

Mollie has been keeping loose tabs on what's been going on around her, but she's tuned out most of Rachel and Brittany's conversation, so she asks for more information. "The storycards for today don't mention anything about building in a registration link," Brittany says, "but we know one needs to be added into the design. The question is, do we need to create a separate card for this task?"

Mollie: "How long do you think adding in the registration link would take?"

Brittany: "In theory it shouldn't take more than two hours, and it shouldn't significantly impact our original estimate for project completion."

Mollie: "Okay, so my suggestion is, go ahead and include the registration link. After the Show & Tell, do a design analysis, find any other gaps in the current storycards, and write storycards for those gaps. If there are several smaller tasks, combine them into one card."

Brittany and Rachel look at each other and nod in agreement. "Thanks, Mollie," Rachel says, "we thought that might be the way to go, but having you think that too makes me feel much better."

Mollie's phone rings. She gives Rachel and Brittany a little wave and picks up. It's the lead tracking company calling back.

"Hello, this is Mollie. Thank you for calling me back. I have my username and password written down, but when I go to log into your website it tells me I have the wrong information. Can you give me the username and password you have in your system?"

As the customer service rep gives her the information, Mollie has an "Aha" moment. The issue isn't that she had blended her personal and work emails on the forms as she originally thought, but rather there was a transcribing issue on the lead-tracking provider's end of things. Whoever did the data entry for the Menlo account transposed the "io" in Innovations to "oi" when entering Mollie's email into the system.

"I was feeling like a 'stupid user,'" Mollie tells the rep, with a chuckle. "Stupid user" is a term widely used by developers to account for a problem with the software they've created. If the software isn't working properly, it's the stupid user's fault. It can't be because of some flaw in the developer's brilliant design. Needless to say, Menlo doesn't believe in "stupid users." Their mission, their whole process, is about making sure that their software designs accommodate the real experience of the intended user.

"Thank you and have a nice day," Mollie says. Going back to the app with the correct username, she pokes around for a few minutes before realizing that she's going to need a computer with Excel, which the HTA box she's using doesn't have. This task is turning out to be a bit of a pain.

Mollie can, and will eventually have to, move to a computer that's loaded with Excel. But for the moment, she turns to another task. On each day of the trade show, she and David had run a raffle, offering something different each day, such as a signed copy of *Joy, Inc.* or a tour of Menlo. Mollie thinks that rather than picking raffle winners out of a hat at random, the Liberty team should choose cards submitted by attendees who looked like real leads. After all, that was the whole point of their having a booth at the show. But she decides to see what Anna thinks, and walks over to Anna's table. "Hey Anna, I have two questions for you. One: What do we want to do about the raffle?" She holds up a couple of plastic bags with business cards in them.

"Do we want to *not* be completely random?" Anna asks, echoing Mollie's thoughts.

"That's what I'm thinking too. Okay, great. One other question. Is there a PM computer free that I can use? I need Excel to download the leads from the database."

"It looks like there's one right over there," Anna said and pointed to a group of computers just in front of the Work Authorization Board.

"Okay, thank you. Before I go, I have to tell you what happened to me when I went to log into the lead-tracking system this morning. I tried using the username and password I had written down and it didn't work. I thought I was a stupid user, that I had messed up my email and that it was my fault, but after I talked to their customer service rep, it turns out that it was a data entry issue on their end. They misspelled Menlo Innovations."

"Wow," Anna says. "I get that wrong username or password message all the time, and I can't think of a time when it wasn't my fault."

Mollie heads back to her table, picks up her purse, her phone, and her notes from the trade show, and logs out of the computer. Then she crosses the room to the Work Authorization Board, sits down at the PM computer Anna pointed to, and lays out her notes. After what seems like a messy start to her day, she can finally get on with downloading the contact info for the leads from the trade show. And then, from an electronic dartboard hanging on the wall next to the Work Authorization Board, an alarm goes off.

9:30 AM
JARON VOGEL & GRANT CARLISLE
Developers

> *Simplicity—the art of maximizing the amount of work not done—is essential.*
> -- A Principle behind the Agile Manifesto

Jaron and Grant are still working on Night Sky. Having fixed the data formatting—or at least gotten it to a place where it will produce clear feedback from the client's project manager on the lines or no lines issue—they can move on to the second item on the storycard. This involves implementing a sorting function to organize hospital patient data by name, age, room number, or physician name.

Before they get going, Grant asks if they also should provide the ability to sort a small subset of the data, a feature that would certainly be useful for the client to have. Jaron scratches at the stubble on his chin and reads the storycard again before responding: "That's not on the card, and I don't recall seeing it on any of the other cards for this week, so I think we should just stick with what's here." He taps the storycard on the table in front of him.

"I don't know," Grant says, "I think that this feature would really make the client happy and I don't think it would take us that long to do. We could add a storycard for it and work on it later." Jaron nods and replies, "Well, you have to ask yourself 'what hat am I wearing?' If you're wearing a dev hat, then you do what's simplest while matching what the client asked for. If you're wearing an HTA hat, then you focus on the user and their needs as much as possible."

Grant looks up at the ceiling and takes a long minute to think before he looks back at Jaron and says, "Nope, you're right. They didn't ask for it. Sorry."

From the adjacent table, where Al and Travis are working on their own storycard, Al looks over and says with a smile, "That's why we pair!"

This whole, brief set of interactions—the open dialogue, the lack of defensiveness, the lack of an implied hierarchy—speaks volumes about how the Menlo culture operates and how it gets transmitted to newcomers.

✦ ✦ ✦ ✦ ✦

Over the next 20 minutes, Jaron and Grant pick up quite a bit of momentum. As the sort function begins to take shape, their work has a continuous flow: Jaron writes a test, Grant writes code to pass it, and then they switch roles. But then, Grant sits back from the keyboard and points at the computer screen. "This piece of code seems overly complicated, don't you think?"

Jaron: "Are you saying that because it's more code? It gives us more control over what we look at and how we sort."

Grant: "Yeah, but do we need that much control for what we're doing? I mean, this way it requires three lines and requires understanding the complexity in that middle line. Let me ask you this, would you expect it to be this long?"

Jaron: "That's a fair point." He leans forward and squints at the several lines of code in question. "Maybe there's a default we can use."

With that, the two begin looking for a way to implement the sort without making it so complicated. While Grant works the keyboard and the mouse, Jaron leans in, reading what's happening on the screen, occasionally asking a question. Before long, the messy three lines of code they started with has been reduced to one. Grant sits back while Jaron leans forward to read over the line more carefully.

> **ENABLE COLLABORATIVE EMERGENCE**
> True collaboration requires an environment of free and open discussion that enables people to synergistically create something that no one of them would have come up with on their own.

"I think this looks good, but we're sorting on the wrong thing," as he points at the line that Grant has just completed. "But since we're working on a simple case, maybe we can sort this way. May I?" Jaron motions to the keyboard and Grant scoots over slightly so Jaron can access the keyboard and mouse.

Quickly, Jaron turns Grant's solution into two lines of code, and after another five minutes of testing, the two partners agree that the sort now works the way it should. Bravo for *collaborative emergence*!

Grant stands up and stretches his arms. Jaron takes a swig from his water bottle. And just then the dartboard alarm goes off.

9:43 AM
KEVIN F. & RICK COPPERSMITH
Developers

Back at their worktable, Kevin and Rick have decompressed from the long Quail stand up. They've gone over Rick's notes and talked about what they and the client team committed to. The stand up went well, especially given the data entry problem that had crashed the program on Wednesday.

> Developers have as many definitions for 'done' as Eskimos have words for 'snow.'
>
> –Richard Sheridan

A poster on the wall at Menlo Innovations

Now it's time to "orange dot" a piece of code they finished yesterday. When a programming pair thinks a piece of code is good to go, another pair steps in for a review. So now, Kevin and Rick have been joined by two other programmers, who've come over to their table and are now sitting at the computer running the new code. Questions go back and forth. Menlonians ask a lot of questions.

And then the alarm goes off.

9:58 AM
MATT SCHOLAND & ART KLEIN
Quality Advocates

Matt and Art are discussing whether to go to the 1:00 pm Quail Show & Tell or keep working on their storycards. They agree that they probably won't have involvement at the Show & Tell, but they'll check with Michelle, the Project Manager.

They turn to Card 347, focused on changing the payment history date from past to present. Should be pretty straightforward.

And the alarm goes off.

10:00 AM
ALL MENLONIANS
Daily Stand Up

At 10:00, the dartboard clock on the Work Authorization Board sets off an alarm and everybody in the office, including any visitors and clients, heads for the middle of the floor for the Daily Stand Up.

The idea for the Daily Stand Up goes back to the Java Factory at Interface Systems, in the day pre-Menlo days. Someone suggested that a brief meeting every day, with everyone standing up, might be a good way to keep everyone informed about what was going on without wasting as much time as the typical update meeting. The Stand Up was set for 4:00, but when that time rolled around, everybody just kept working at their own table. No Stand Up.

The same thing happened the next day, and the next. But then, on the fourth day of this experiment, a developer named Tim got hold of an electronic dartboard sitting in a closet, tinkered with its programmable alarm, and hung the thing on a wall. And then, at 4:00, a soft ringing noise started emanating from the dartboard. As everyone looked around to see what was going on, Tim stood up and announced that it was "time for Daily Stand Up." Rather slowly, everyone got up and headed for the middle of the room, and thus, the Daily Stand Up was born.

As Rich explains, the next inflection point came later, when people started treating Stand Up as "stand around." Instead of paying attention to who was supposed to have the floor at a given moment, people began having side conversations. Drawing on elementary school practice, someone suggested using a "talking stick" to ensure that only one person had the floor (although at the time, the artifact was actually a "talking coffee cup").

Another "inflection point" came when people began to effectively monopolize the Stand Up, pulling out lists of items (much to the dismay of everyone else) and, coffee cup in hand, droning on, defeating the idea that the Stand Up was to be brief. The solution: replace the coffee cup with a dumbbell. From then on, you could only talk when you had the talking dumbbell and for as long as you could hold it straight out in front of your body. The result? Daily Stand Up rapidly shrunk back down to its intended length of around 11 minutes.

Over time, the talking dumbbell was replaced by various random objects. Then, one day, designated at Menlo as "crazy hat day," someone showed up wearing a plastic Viking helmet. They used it as a talking stick that day and found that it was perfect for *pairs*, who could each hold one horn of the helmet as they gave their update. The principles of pair programming could even be rolled into Daily Stand Up! It was almost poetic, really.

✦ ✦ ✦ ✦ ✦

Today's Stand Up is pretty typical. There's some joking around and light banter while those already in the circle wait for those on their way. James and Andrew Muyanja, an HTA, get a laugh out of the fact that they're wearing matching sneakers.

Finally, everyone—Menlonians, Menlo babies, Menlo dogs, Menlo visitors—is in the circle. There's even a remote attendee. Randy has worked for Menlo for several years, but when his partner got a good job in Russia last year, Randy moved there too. The company didn't want to lose him, so for six months he's been working from Moscow. Although Menlonians have sometimes worked temporarily from a remote location, just how well remote work fits into Menlo's culture remains an open question.

The experiment with Randy ups the ante, but at least so far, it seems to be running smoothly.

A pair of Project Shropshire developers get the meeting started. They introduce themselves (everyone, including Rich, introduces himself or herself at Stand Up), and then, each holding a horn on a toy Viking helmet, they announce that they're on the final stretch of their project. The whole group applauds, and the helmet gets passed along the circle. One pair tells a story about an issue they ran into on their project the day before. Another pair reports that "Ethiopia has just elected its first woman president." (People clapped. McTavish barked.) Someone announces that the Halloween Costume Contest is on for next Wednesday and encourages everyone to participate.

When the helmet gets to James, he announces, gesturing toward his shoes, that "for whatever reason," Andrew is trying to copy his style. Jane announces that she's working on payroll and that she plans to leave the office early since she's not feeling well. All the while, Egon, one of the Menlo dogs, rolls around in the middle of the circle. The helmet keeps moving.

James remembers that he has another announcement, so, following Daily Stand Up protocol, he moves to another spot in the circle ahead of the Viking helmet. This time when the helmet gets to him, he says, "Hi. I'm still James," which produces a smattering of chuckles. Then he adds, "I'd like to go ahead and give the award for the most team spirit to Team Night Sky for their fantastic team uniforms." There's a brief moment of confusion before the circle erupts with laughter as everyone turned away from James to see that not only had Jaron and Grant come to work in flannel shirts this morning, but in fact, Al and Travis are clad in flannel as well.

The helmet having made the complete circle, it's time to disband. As always, the last person to speak ends the meeting with an iconic phrase from a famous '80s tv show, Hill Street Blues: "Be careful out there."

Time elapsed: 11 minutes!

Daily Stand Up

10:15 AM
JARON VOGEL & GRANT CARLISLE
Developers

Back at their table, Jaron fishes his phone from his pocket and says, "I'm going to log my time before we get back into it."

"Good idea," Grant replies, as he pulls out his own phone.

Menlo employees bill their time in 15-minute increments. Ideally, developers end up with 32 hours per week of time billed directly to working on specific storycards for client projects. Show & Tells, and Kick-Off meetings, plus Daily Stand Ups and other such activities are baked into estimated 40-hour work weeks. Menlonians have come up with a slew of different ways to keep track of their billable time as they go through a day. Some have built apps that allow them to do that on an on-going basis. Every time they start work on a new storycard, they just input the project name and storycard number, and then the app records the time spent. Others, like Jaron and Grant, use simple spreadsheets that they fill out every hour or so. Some go old school and use pencil and paper to keep track as they switch tasks during the day.

In most companies in which employees need to track billable time, a particular method/tool would be mandated for doing so. Everyone would do it the same way. What does it say about Menlo's culture that Menlonians are free to use whatever method works best for them?

10:15 AM
KEVIN F. & RICK COPPERSMITH
Developers

Kevin and Rick are back at their table. Although they already talked through the Quail client stand up, before their earlier orange dot code review, they come back to it now. Using Rick's notes to refresh their memories, they go back and forth, unpacking every aspect of the meeting and once more reviewing Wednesday's data input "emergency" and the commitments they've made to the client. Finally, they sit down to write a follow-up email to the client.

On his way out for a walk, with Marilyn in tow, McTavish stops by their table for a scratch.

Rick believes that pair programming is useful and effective about 85% of the time, with the other 15% of the time difficult to gauge. For example, does it really take two people to write an email? Maybe not. You could ask the same question about inputting data into an application. But somewhere in between individual efficiency and the tendency for mistakes to be overlooked when there's only one set of eyes involved, that's where you get that 85% benefit of working in pairs.

At the moment, Rick is on the keyboard. As he types, he reads, and Kevin jumps in with comments and suggested edits. This is the way pair programming is supposed to work. It's not appreciably more time consuming than having one person do it alone, but the final product is almost always better.

The email is done. Kevin and Rick both read it through one more time. "Good to go?" Rick asks. "Yep," Kevin replies. Rick sends the email off to the client using Slack, an interactive tool for sharing files, tagging specific people, making calls, and sharing a desktop. Slack allows Menlonian developers to keep in close touch with their clients.

Following up on the client stand up, Rick and Kevin now need to write some new storycards. After only four months at Menlo, Kevin still hasn't fully mastered the art of storycard writing, so this is a teaching moment. Kevin starts by talking about what the card should include; Rick comes back with immediate feedback. When they've agreed on what the card needs to say, including specific language, Kevin finally starts writing.

The last element of the storycard is the time estimate, which Rick and Kevin have pegged at eight hours. They know, and the client knows, that these estimates are just that, estimates. The actual time spent might be more or less, but the estimate provides some useful guidelines. Given all the experience accumulated over the years, Menlo estimates also tend to be pretty close.

With the storycard finished, Kevin calls over to ask a nearby programming pair to look it over. Title, task summary, and estimate: they get a thumbs up. Next, they ask Michelle, the Quail project manager, to come over for a final check. She asks a few questions, then gives her okay.

Back to that programming pair that just reviewed the new storycard. Kevin and Rick take their colleagues through that morning's client stand up. There's some tech-speak about auto matching. They separate once more.

The morning is moving along nicely. Kevin and Rick send an "orange dot" to the client. Code awaiting code review. Code review from the client, specifically. Almost immediately, the client comes back with a thumbs up; the card is now ready for a QA review. Pair programmers, to pair programmers, to clients, to QAs—it's a smooth process, aimed at making sure that the final product delivers what the client needs.

Daily Stand Ups, High Speed Voice Technology ("Hey Menlo!"), Show & Tells, pair programming, etc.—practices, processes, artifacts, and rituals, all driven by and reinforcing the underlying Menlo culture—all contributing to the sense of joy.

10:15 AM
BRITTANY MORTON & RACHEL CLEVELAND
High-Tech Anthropologists

Working in pairs, at least at Menlo, seems to foster accountability for how time gets spent. As they keep working on the prep for their upcoming Show & Tell, Rachel and Brittany crack some jokes about cats, and talk a little about their personal lives. ("I didn't sleep well last night," Rachel says as she works on a PowerPoint slide.) But they're always quickly back to business.

Part of the HTAs' job is to create mockups of how the software should actually display, based on their observations and interviews with the end users. The mockups are run by 3-5 users who are representative of the client company's primary persona. The mockups intentionally look more like rough sketches than refined designs. They lack the design embellishments, color schemes, branding and so on that will eventually be added to the final product. Experience has shown that users are more likely to give negative feedback on a design that appears to not yet be finished, and the goal is to get at potential design flaws before the project gets too far along.

Clients, as opposed to end users, typically don't see the mockups until the end users have reviewed them and the mockups have been revised accordingly. This helps ensure that the primary persona's needs take precedence. In the case of the Quail project, Rachel and Brittany will be presenting the mockups to the client at today's Show & Tell.

At Menlo, more input is almost always better than less. Brittany and Rachel sit back from their work and call over to Andrew and Eric, two other HTAs, to ask if they can take a look. Andrew and Eric stop what they're doing and come over. Because HTAs work on several projects at a time, they tend to be less invested in imposing their own views on any particular design. That makes it much easier to stay open to feedback from other Menlonians, and more importantly, to let the design flow from the users' needs.

Andrew and Eric make a few comments but agree that the mockups look good. They head back to their table, and Brittany calls out to Mollie, who has been involved on and off with Quail from the beginning. More input is better than less.

10:15 AM
JANE, MICHELLE, LISA, KEALY, ANNA, MOLLIE, JAMES
Experience Team

After Daily Stand Up, the Experience team gets together for its own mini stand up. The job of the Experience team is to get the Menlo story out to people who are curious about the company's values, practices, and results, but aren't necessarily prospective clients. They plan and execute presentations on Menlo-related topics to community organizations, group tours of the Menlo Factory, Menlo workshops, and a variety of other events.

The problem at hand is that they have committed to three presentations on three different Menlo-related topics at three different venues, all within the same, rapidly approaching time frame. But as the planning has gotten underway, the team has started to grow concerned about whether or not they have the bandwidth to successful carry out these commitments. Hence the mini stand up.

In many organizations, this kind of situation would devolve quickly into finger pointing, as in "Here's another fine mess you've got me into," to quote from the classic Laurel and Hardy routine. At Menlo, however, when people are dealing with difficult workload-related issues, the go-to phrase is "Do you have time for this?"

When Menlonians ask one another "Do you have time for this?" the question is asked, and taken, seriously. It

> **TREAT INDIVIDUALS WITH RESPECT**
>
> In high-performing organizational cultures, people are seen—and they see themselves—not as "human resources" but as actors with the power to make a difference, to contribute, and to grow.

comes down to treating one another with respect, to really caring about one another, and about being trustworthy—only committing to what you can really accomplish and then doing whatever you've committed to do. These are all axioms, values, and principles critical to the Menlo culture, and even when not explicitly invoked, they are taken seriously.

So, the question for the Experience team seems to be: Does the team, individually and collectively, have time for this? Not honoring the commitments they've made is hardly a good solution, even if those commitments were made somewhat unwisely. But if the commitments are to be honored, who exactly has the time and will commit to doing the necessary work?

James says, "Feels as if the team is already spread a bit too thin for everyone's liking. We need to decide what the best way to proceed is." Nods of agreement.

The three talks cover quite different topics: the importance of Persona Mapping; what is involved in Quality Advocacy; and how an organization can remove fear from its culture. "Are there past talks or presentations we can pull material from? I know we've given similar talks in the past," James asks.

"There may be, but we'll have to look," Anna replies. "And there may not be an exact match with any of the topics."

Mollie says, "I can do a 30-minute brain dump today for the speeches and have that ready for Monday." Mollie was involved in the initial discussions of the three events.

Kealy says, "I feel a little disappointed because I have been wanting to contribute more to projects like this and feel like I'm being overlooked. Giving this kind of talk is one of my professional development goals."

"If that's the case, would you be willing to work outside of normal hours to get it done?" James asks, to make sure Kealy is fully aware of what she's committing to. "Yes," Kealy replies. Kealy, apparently, *does* have time for this. Several people in the circle give her a smile or thumbs up.

"Okay, then, can we pick this up again on Monday?" James asks. More nods and the group breaks up. Although there seems to be some positive movement, the problem is hardly solved. It still seems clear that putting the three presentations together is likely to put a strain on people's time, Kealy's commitment notwithstanding.

Nobody, certainly not Rich Sheridan, has ever said that joy is always easy.

10:20 AM
GUINEVERE PROJECT TEAM:
ANGELINA FAHS, GIOVANNI STURLA, HELEN HAGOS, KYLER WILKINS & LISA AND JOSIAH H.
Show & Tell

The Guinevere team is waiting for Carl, the client project manager, who was supposed to be there at 10:00. He's late but nobody is worried. If he'd run into a major problem, he would have called in. The video connection into the client's conference room is up, but no one on the client team has shown up on that end either. Again, nobody on the Menlo team is worried. They just default to working on the project.

As they discuss the project schedule and various coding issues, the Menlonians also make small talk about their families, their weekend plans, etc. Someone asks Lisa, the project manager, how six-month-old Josiah is feeling. He's been running a fever for the past couple of days, so he hasn't been in the office, but today he's back. At the moment he's sleeping soundly in a baby pack on Lisa's chest. Lisa remarks that he's fine, but that he's getting so big and active that it will soon be time for him to transition to day care. Having Josiah in the office has been another great example of Menlo's successful "bring your baby to work" policy, and it's clear that he'll be missed.

Show & Tells usually are intended to cover the one or two weeks of work just completed, and the next one or two weeks of work. For Guinevere, however, Carl wants to be part of the project's long-term planning, so this Show & Tell will lay out the entire timeline for the project. On tables set out in the conference room, Lisa has laid out differently colored sheets of paper, two sheets for each iteration. Taken together, the layout provides an easy-to-follow visual representation of the overall project flow.

Each sheet is folded according to the time estimate for the work the sheet covers: an unfolded sheet for 32 hours, a sheet folded in half 16 hours, in quarters 8 hours. Menlo regularly uses this visual representation of estimated time in Show & Tells; it has proven to be very useful in helping clients make decisions about what work they want done in the immediate future. But this approach has never been used to lay out the entire scope of a complex project. In Menlo speak, in using the approach in this way, Lisa is "running the experiment."

At 10:25 the client project team calls in. Quick greetings are exchanged. The meeting can't really start until Carl shows up, but the Menlo team does share the fact that a process that had been running smoothly had suddenly started to crash. The problem seems to be that some data in an unexpected format had been uploaded by the Menlo team. Now that they know, the problem has been resolved. Someone on the client end says, "Thanks for letting us know. Your transparency is appreciated."

Carl still hasn't appeared, so Helen asks a question about the workflow. The client project team seems to appreciate her thinking, and someone says, "Thank you for thinking of that." After some additional back and forth, someone on the client team suggests that Carl might have had a dental appointment that morning. While it seems strange that he wouldn't have let everyone know, both teams decide to move ahead without him. They'll bring him up to speed when he gets in.

"Hey, Menlo," someone from the client team says. He holds up a plastic Viking helmet. "Are we planning on a Stand Up today?" The Menlo team breaks out laughing at this adoption of their ritual. The client team passes the helmet around, with pairs of team members giving brief updates on the work they've been doing. When they're done, the last pair pretends to hand the helmet over to the Menlo team, 50 miles away. More laughter.

10:20 AM
MATT SCHOLAND & ART KLEIN
Quality Advocates

> *The most efficient and effective method of conveying information to and within a development team is face-to-face conversation.*
> -- A Principle behind the Agile Manifesto

Matt and Art are back working on the Quail project. Storycard 306. They previously red-dotted this card, and the developers then re-worked the code. But by the time that work was done, Matt and Art were working with other partners, so the card was not QA reviewed again. Generally, it's better for the original QA pair to re-review a card they've red-dotted, to make sure they are in agreement that it should get a green dot.

Now, Matt and Art conduct the re-review and find that the issue has been fixed. They give it a green dot.

Then it's on to Card 22, focused on a feature of the software that's supposed to automatically match payments to invoices. The client, a co-op, is moving its invoice payment system off a mainframe, and wants to automate the system in the process. When they start to test the matching feature, it seems to work but they decide that they need to understand the code better and possibly test it further.

Looking for help, they turn to two developers at the next table. After listening to Matt and Art quickly explain their issue, the developers call another pair of developers to come over and join the discussion, which then goes on for five minutes or so. Then the group decides to bring Michelle, the project manager, into the discussion. Michelle is sitting only about 10 steps away, and she comes right over. More discussion. Finally, with the entire group peering at the same computer, two of the developers write a storycard to resolve the issue Matt and Art have uncovered. Estimated time: eight hours.

This whole process—from the initial discussion between the QAs and developers, on to further discussion involving the other developers and then the PM, and finally on to defining the task necessary to solve a potential problem—this whole process took just under half an hour!

The group breaks up and Matt and Art turn to the next item on their agenda. They need to develop a more realistic set of test data so they can more rigorously test the automatic payment/invoice matching function. Again, they call on their developer colleagues for help.

Matt leans over and asks a developer at the next table, "Hey Michael, if we're inserting an invoice and want it to be blank instead of null, what do we add?" Michael tells them how he would do it. Matt and Art talk, and Matt turns to Michael with more questions, asking about various scenarios. Finally, with Michael's guidance, they've developed the test data they need.

All of this back-and-forth between the QAs and developers happens at high speed. They don't send emails to each other and wait for a response. Instead, they call out to one another and talk to one another face-to-face, the interactions taking minutes (sometimes even just seconds). It's all very focused, and it all happens fast.

As the morning continues, Matt takes extensive notes on potential issues and what they might want to communicate to developers and PMs. Art works differently, taking notes on only a few of the most important issues. Is this because he can count on Matt to cover the others? Is this another advantage of pairing?

In the same way, Matt also fills in his timecard throughout the day. Since Art has been with Matt all week, his plan is just to make a copy of Matt's timecard at the end of the day. Let's hear it for pairing!

10:20AM
JANE COLLINS
Project Manager

After the Experience team stand up, Jane and James walk to a nearby table where Harry and his programming partner are talking about the piece of code they've been working on. Jane asks Harry if he has a couple of minutes to talk about a job applicant the three of them have been working with. Harry's partner says, "Go ahead. I'll go grab some coffee while you talk to these guys."

> The aim of an argument or discussion should not be victory, but progress.
>
> —Joseph Joubert

A poster on the wall at Menlo Innovations

Harry quickly summarizes the areas of strength and weakness he's observed while pairing with the applicant on and off during the applicant's three-week trial period. When he mentions areas where the applicant could use improvement, James asks him, "Have you given him this feedback?"

"Probably not as completely as I need to," Harry replies. "I'll make sure to do that the next time I meet with him."

Jane: "What can we do to help him?" It's clear that the applicant has shown great talent as a programmer, and if the team does offer him a position at Menlo, they would all like to help him succeed. Very briefly the three of them brainstorm how he can demonstrate that he would be a good cultural fit.

The conversation they're having is taking place right near the "Levels Poster" on the back wall of the Factory, and that prompts a brief, final discussion of what level the applicant might start at, if he does come on board.

As the Levels Poster displays, there are several different job categories at Menlo: Associate, Consultant, Senior Consultant, Principal, and a new one, Staff. Within each category there are three or four different "levels," each with a different pay grade. The Levels Poster is also covered with small sticky-notes, one for each Menlonian (as indicated by their initials). The placement of each sticky-note on the poster corresponds to that employee's level and pay grade.

In other words, for someone at Menlo to know what their co-workers are paid, they just have to look over at the Levels Poster. Talk about transparency!

Not surprisingly, there's a story behind the Levels Poster. After three days of exploring Menlo and its culture, a journalist writing an article for *NY Magazine* remarked, "Sure, you guys are a bit weird, but it's not like you're putting everyone's pay on the wall." Which, not surprisingly, made people think, "Hey, that's a good idea! Maybe we *should* put everyone's pay on the wall!"

Like so many other things at Menlo, the Levels Poster reflects what Rich describes as a "whimsically irreverent" attitude toward traditional management practices. After all, if the Menlonians are going to be making hiring, firing, and promotion decisions, they'll need to know—and they'll inevitably learn—one another's status and pay grade. It wouldn't make sense to withhold that information from a hiring manager in a traditional organization, so why would you withhold it from the roughly fifty hiring managers at Menlo? And given that, why not just go all the way and post all that information on the wall?

✦ ✦ ✦ ✦ ✦

As the conversation winds down, Jane shifts it away from the applicant they've been discussing. She points to the Levels Poster, to the recently added "Staff" column. The idea behind adding this new job category was to be more inclusive of people who work at Menlo, but not as "regular" employees—like student interns or people who come in at night to clean the office.

Jane, however, says, "The levels wall suggests that any Menlo employee can move up over time, from one level to another, right? But that's not really the case for the people in the Staff column. They can't move up in the same way. And I think it's actually cruel to suggest to people they can move up when we have no intention of doing that."

"I hadn't thought of that," James replies. "That's good feedback." The Staff column on the Levels Poster is still in its infancy; it's another experiment, and honest feedback—both positive and negative—is the only way to determine if it's right for Menlo or not. Jane will need to add her feedback to the discussion.

10:30 AM
MOLLIE CALLAHAN
High-Tech Anthropologist

With so much on her plate, Mollie puts together a mini-Kanban to help keep herself on track. The basic structure of a Kanban is a three- or-four-column table visually depicting progress on a set of tasks. Developers, for example, might use a four-column Kanban to track the progress of various tasks from *pending* to *analysis* to *development* to *test*, with the appropriate columns sub-divided into *doing* and *done*. It's a simple way for project team members to see at a glance where things stand.

For Mollie's purposes right now, she sets up a simple two-column table on a large sheet of paper. In the *To Do* column she lists her various tasks for the day, broken down into sub-tasks, each sub-task labeled on a small sticky-note. As she completes each sub-task, she'll move the sticky-note into the adjacent *Done* column.

With her Kanban filled in, and her day better organized, Mollie turns back to following up on the trade show leads. Since the whole goal of Liberty is to fill the pipeline, she is primarily focused on "hot" leads, individuals who came to the Menlo booth with explicit asks, especially those who came with possible work for Menlo.

She starts by downloading the contact information from the lead tracking app into an Excel file, then converts the numerical rating of each lead's level of interest into a brief summary, based on a key she and David created. Doing so will help her follow up on the leads, but it will also provide valuable information for anyone who later works on Liberty.

Once she has the data entered into the Excel spreadsheet, she hits "print" and heads over to the printer to pick up the hard copy. As she leaves her table, Brittany and Rachel approach her with another Cherubim question.

10:30 AM
JANE COLLINS
Project Manager

When Anna and Jane get back to their tables, an HTA approaches them to discuss some changes she's made to a poster designed to illustrate the HTA process to clients. To many potential clients, Menlo's HTA process can be very confusing. To help explain the process, the new poster breaks it into steps—*interviews and observations, persona mapping, conceptual designs, iterative design, and finalized designs*—with cartoon clip art illustrating what each step entails.

Jane and Anna give the clip art an enthusiastic thumbs up. Then the discussion moves on to how it might be possible to provide even more detail and insight into the HTA process. Perhaps the poster might include some sample HTA "process cards." Menlo has tried this before, without much success, but instead of just dismissing the idea, Jane and Anna discuss why it hasn't worked in the past and how it might be done differently this time.

Finally, they turn to the bigger issue. Even with changes such as those they have been discussing, what exactly will this poster communicate to clients? Will it really help them understand the role of HTAs in Menlo's software development process? And at a more tactical level, how might Menlo enlist clients in helping them come up with an optimal solution?

Anna and the HTA go on talking. Jane walks back to her table to continue payroll.

10:30 AM
GUINEVERE PROJECT TEAM:
ANGELINA FAHS, GIOVANNI STURLA, HELEN HAGOS, KYLER WILKINS & LISA AND JOSIAH H.
Show & Tell

Carl hurries into the conference room ten minutes after the Show & Tell has started. Right on his heels, James Goebel, Menlo's co-founder and COO, also joins the meeting. "Wow," Carl says with a smile, "I must be special." "I can't sit in on all of these meetings," James replies. "I only do it for the clients I like." There's laughter in the room and, via the video link, from the client team.

The fact is that Carl is indeed special. He created the earlier version of Guinevere, which is a cash/event/accounting program targeted at small non-profit organizations that are loosely connected to national organizations. Guinevere, which comes customized to meet the specific needs of each client, makes it easier for them to manage the finances of the events they put on without requiring high levels of skill in cash management. Carl's version has already achieved considerable success, but even with his own development team working on the project, he's retained Menlo to help take his "baby" to the next level.

As for James' presence at the Show & Tell, that's certainly not just because he likes Carl. Given his profound sense of ownership for Guinevere, Carl has asked that the meeting cover the long-range plan for the project, and Lisa and the team are ready to take him through that plan. But even with the best long-range planning, every project—especially a complex project like Guinevere—is subject to uncertainties that can result in unforeseen cost and/or delays. James is at the meeting to discuss with Carl the tradeoffs that may be involved as the project rolls out.

But there's another reason for James' attendance. The Menlo team has serious concerns about the programming ability of one of the developers on Carl's team, and on a collaborative project like this one, less-than-stellar performance on the client's end could affect what the Menlo team can accomplish. The team has asked James to address the issue with Carl, one-on-one.

The mood in the room is relaxed. When Carl comments on the bottles of antibacterial lotion on various tables, someone jokes that "Colds spread faster at Menlo because we work in pairs." With everyone working together to get the right result, and with Menlo's process specifically designed to create a truly collaborative environment, Menlo projects often result in friendly connections between Menlonians and their clients.

The friendly atmosphere is especially helpful to Angelina, the young developer driving today's Show & Tell. She doesn't seem to be nervous, even with the company co-founder and the client project manager in the room.

The meeting migrates into some nitty-gritty technical details. Carl calls everyone's attention to storycard #366. "When we get to this, make sure there's a drop down to go across departments." The Menlonians scribble notes on that.

Giovanni raises a question about card #226, which the team dealt with earlier today. "We had this card as red. But this morning we figured out that the program crashed because of a next integer problem. We fixed that problem, so now should we green dot the card?" The question leads to a discussion of different ways the fix they put in can be tested to be sure it works. At one point, James comments that, "The program should behave this way, but there is a step up in the cost to complete the coding. It's a business trade-off problem. We need to make bad things happen, so we'll know if it's the software or the environment that's the problem." A riff on the theme of uncertainties, timelines, and budgets.

Ultimately, the Menlo team and the client team both agree to test and re-test this piece of the code, as well as various environmental factors, both being open to the possibility that it's something they did that caused the problem. In a show of the trust and collaboration that exists between the client and Menlo, it's agreed that the responsibility for paying for the time spent correcting the problem will be determined by the outcome of the tests on both sides.

10:40 AM
JANE COLLINS
Project Manager

> Simple can be harder than complex: You have to work hard to get your thinking clean to make it simple.
>
> —Steve Jobs

A poster on the wall at Menlo Innovations

In the last half hour, Jane has participated in a Menlo Experience discussion, a hiring discussion, a discussion regarding a new Menlo experiment (the Staff "level" issue), and an effort to better communicate the HTA process to clients. In most organizations, each of these discussions could well have involved its own sit-down meeting. To set up the meeting, time would have been spent coordinating calendars, booking a conference room, sending out an agenda, and so on. People would have drifted in for the meeting in dribbles, and more time would have been spent in chitchat waiting for others to arrive.

And as for the meeting itself: at least half an hour, and probably more. For the four meetings, you're talking about a minimum of two hours of actual meeting time, not counting the time to make and confirm the arrangements. Would more have been accomplished? What do you think?

✦ ✦ ✦ ✦ ✦

Back at work on the payroll, Jane prints off an Excel spreadsheet showing all the levels and corresponding pay grades. On her way back from the printer, she stops to give a copy to Harry, the developer she and James talked to about the job applicant they've been assessing. The levels information should help inform his thinking about the best level the candidate might start at.

✦ ✦ ✦ ✦ ✦

Jane has been thinking about what is going to happen as the company grows. With a significantly larger number of Menlonians, many of the administrative tasks she handles will become impossible to manage in the same way. What new systems will be needed? Will they need to hire an HR person to manage this stuff? How would bringing in an HR manager affect Menlo's commitment to having the whole team involved in hiring, firing, promotions, performance feedback, and other employee-related issues? She remembers how badly Menlo's experiment with a professional salesperson turned out.

On the flip side, Jane also knows that a trained, experienced HR person *could* really help her deal with a variety of issues. One of her tasks for today, for example, is to establish Menlo as a company sponsor for a co-worker's visa, something she has never worked on before. And with the present size of its workforce, Menlo would soon be subject to different employment standards under the Affordable Care Act (ACA) and the Family and Medical Leave Act (FMLA). Jane has a key role to play in making sure Menlo meets those requirements, even though she has never dealt with them before. "We, and that certainly includes me," she says, "are learning as we go with this. But that's how we do it."

✦ ✦ ✦ ✦ ✦

Jane goes back through the timesheets, the Excel spreadsheet, and QuickBooks once more to make sure all the hours match. Then she sends an email to each Menlonian informing them of their PTO (Paid Time Off) balance. She started this practice in response to feedback from several Menlonians who said they would appreciate more frequent PTO updates. She's been asked by various developers on the team why she doesn't set up a script to send out all the PTO emails, but she prefers to send them individually. She has a template: "Hi, [insert first name], as of October 28, 2018 your PTO balance is [insert balance]. Have a good weekend, Jane."

She knows that her whole payroll process is manual, home-grown, and somewhat time-consuming, but at least for now, that's not necessarily a bad thing. "Sending out emails individually, updating Excel manually, inputting hours into QuickBooks from printed timesheets seems like the hard way, but you're more likely to get it right."

Strange as it may seem for a software development shop that takes special pride in delivering technology solutions that really work the way they're supposed to work, Menlo's (unspoken) belief seems to be that in many cases, the best way to end human suffering as it relates to technology is just to use less technology.

10:45 AM
RICH SHERIDAN
CEO and Chief Storyteller

On his way back from the SPARK meeting, Rich impulsively gets off the freeway before his usual exit to check out some properties that might, if they're still available in four or five years, be appropriate for Menlo. The Menlo 2027 Vision calls for a workforce of about 200 people, four times the current size, and that would require a move.

Rich and James have talked about what might fill the bill. Maybe something like a former department store they've looked at, with a 70,000-square-foot showroom that would definitely handle 200 Menlonians. Something they could buy and own, not just lease. Something that might allow them to bring along a startup or two, or some of the other Ann Arbor businesses that they have developed an affinity for over the years. Maybe include an onsite daycare center so Menlo babies could graduate to something closer by. Something more permanent.

But do they really want Menlo to get that much bigger?

ZINGERMAN'S DELI

Zingerman's Deli is a local Ann Arbor business that has inspired Rich. Zingerman's is very popular—world famous even!—but as the business took off, Ari Weinzweig, the founder, avoided the obvious opportunity to franchise the deli and expand all over the country. Instead, Zingerman's branched out to other related business that could be clustered in the Ann Arbor area, like a mail order business, and a bakery and a creamery. Rich has taken a number of ideas, including Open Book Financials, from Ari, and Ari has given Rich some guidance on the Menlo 2027 Vision.

Zingerman's Community of Businesses was cited by Bo Burlingham in his book *Small Giants: Companies That Choose to Be Great Instead of Big*. Menlo has also ended up often being on the Small Giants annual list of companies that choose quality over quantity. One of the features of these companies, according to Burlingame, is that the owners have a deep appreciation for the local communities they serve. Their companies have deep and extensive ties to those communities. Rich remembers all the years he worked in corporate IT, when he was located in Ann Arbor but wasn't really present in this amazing town. He's resolved to never let that happen again.

There are certainly potential upsides. A bigger Menlo might be a more financially stable Menlo. More importantly, a bigger Menlo might offer better evidence for the core ideas that Menlo has brought to the world of work. Right now, it's too easy for people to dismiss what they do by saying, "Yeah, sure, that works for 50 people. But would it work for 500? For 5,000?"

The danger, of course, is the challenge that 4x growth might pose to the Menlo culture. The core axioms, values, and principles should be just as valid in a 200-person Menlo—there are bigger companies out there with those same cultural foundations—but what about Menlo's practices and artifacts? Could 200 people hear "Hey Menlo!" without a sound system?

10:52 AM
MOLLIE CALLAHAN
High-Tech Anthropologist

Brittany and Rachael approach Mollie. "We're working on the presentation for this afternoon's Show & Tell and we want you to green dot some slides," Brittany said.

"Sure, let's have a look," Mollie responds. They head back to Rachael and Brittany's table.

"We're thinking of including some notes and annotations on the slides to better explain what we did," Rachael explains, showing Mollie the first of the slides.

"Since you're trying to show the changes you've made to the design, I think it might be a good idea to include before and after titles on the slides over the images that show the changes, as a way to better highlight those changes," Mollie suggests. "As for the notes, maybe you could add them in after the meeting and send them over to the client for review. This might simplify the slides and the presentation. How does that sound to you two?"

"I agree," Rachael replies. "Me too," Brittany adds. "Sounds like a plan. Thanks Mollie."

✦ ✦ ✦ ✦ ✦

Mollie leaves the pair and makes her way back to her table to start reviewing the trade show leads. The leads of most interest are, of course, those where someone had expressed a need for software design and production, followed closely by those interested in Menlo's possibly consulting on company culture and processes.

As Menlo has become increasingly well known for its unique corporate culture, more and more companies have come looking for help in that area. Menlo is more than happy to help reduce unhappiness whenever and wherever it can, and it doesn't hurt that consulting is a growing item in the company's business portfolio.

Back at her table, Mollie works through the list of leads looking for decision makers who expressed an interest in software design or consulting. She's identified three hot leads when Menlo's High Speed Voice Technology fires back up.

11:00 AM
KEVIN F. & RICK COPPERSMITH
Developers

Kevin and Rick stick a purple dot on the storycard they've been working on. This indicates progress somewhere between orange (work completed by the developers) and green (approved by QA and considered "done"). This particular storycard has been reviewed by another set of programmers and sent to the client's Database Administrators and software engineers for review.

Kevin says, "We're running an experiment here."

"Running an experiment" is as foundational to Menlo as pair programming and making mistakes faster. The presence of McTavish and his fellow Menlo canines is the result of a successful experiment: Menlo canines. The same could be said for Josiah, sitting in a baby carrier strapped to his mom: Menlo babies.

In this case, Kevin is referring to their sending *pieces* of code in for review, rather than sending in a more complete product. It's an experiment.

They look up to watch a couple of Menlonians slide a portable wall across the room to the reception area in preparation for their afternoon Show & Tell. At the next table over to their right, Vaughn, who's applying for a developer role, is working with his partner. Today is the final day of Vaughn's three-week trial engagement. Something like 60%-70% of three-week interviewees are ultimately hired on. Has Vaughn proven himself in these past three weeks?

Kevin and Rick turn to a discussion of the "action items" that came out of the morning's client stand up to review the failure the other day of the Quail program. That failure has prompted some discussion as to whether or not it constitutes a "software emergency." As testimony to the quality of its work, part of the Menlo story is that the company "has not had a software emergency since 2004." Should this now be revised? Kevin, for one, doesn't think so. The program crash was the result of something on the client end. Menlo had simply swooped in to help.

As they work, they stop periodically to document their work, each of them entering the information in a notebook. They could use the computer, but they don't. Every time they switch tasks, they jot down the time and who should be charged (according to each 15-minute interval) for their work.

11:06 AM
MOLLIE CALLAHAN
High-Tech Anthropologist

From an adjacent table, Anna asks, "Hey, Mollie! Can you come read this email?"

"Sure." Mollie gets up and walks over to Anna and looks over her shoulder at the email in question. The email is Anna's response to a company that wants to bring a large group to Menlo's upcoming free tour. The tricky part of this particular request is that this company is a potential client, so Anna wants to be especially careful with her response. This is all tied up with Project Liberty, which Mollie is also working on, so she's the perfect person to consult.

"Looks good," Mollie assures Anna, then heads back to her table.

First, she logs her time for the morning. Basically, it's all been spent on Project Liberty. She's had a few brief consults with Brittany and Rachael about Cherubim, but each of those was less than 15 minutes, so they're not billable and she doesn't record them.

Back to following up on those hot tradeshow leads. She pulls up HubSpot, Menlo's CRM system, but then can't remember the log in data, so she walks over to Carol's desk to retrieve the "Super Secret Red Book." This plain red binder contains all kinds of information that Menlonians might need, including log in info for various critical systems. For being "super-secret," the book certainly gets around. Of course, no one takes the label seriously because secrets are not really part of the Menlo culture.

Mollie takes the binder back to her desk and searches through the pages for HubSpot.

11:15 AM
GUINEVERE PROJECT TEAM:
ANGELINA FAHS, GIOVANNI STURLA, HELEN HAGOS, KYLER WILKINS & LISA AND JOSIAH H. AND CLIENTS
Show & Tell

By 11:15 all of the cards for the Show & Tell have been reviewed, so the discussion turns to upcoming cards. Half an hour later, the upcoming cards have been confirmed, and the client team breaks off to grab lunch. This seems like the right time to address the issues that have brought James to the meeting.

"We want to give you some early feedback on a concern we have with one of your developers," James says. His tone is kind and measured. "We're not sure he's got a sufficient skill level for this project."

Carl is not defensive. Indeed, it almost seems as if the feedback isn't exactly unexpected. "I can move him to another pair to get more insight into his capability."

"Okay," said James. That was easy. "We can revisit in two weeks."

The discussion turns to Carl's desire to add more Menlonians to the project. "The constraint," James says, "is availability. We're really busy, and many of our clients are asking us to ramp up their teams."

"Okay," Carl replies. "I understand. I'll look at my end to see if there are any issues that might be impacting my own team's speed. Hit me in the face with a fish if I need it."

James smiles. "Solve the people stuff and get everyone to work together, and everything else takes care of itself."

A Planning Game to confirm storycards to be worked on

11:30 AM
JANE COLLINS
Project Manager

After Jane wraps up payroll, she moves on to a reimbursement to a Menlonian for project-related travel expenses. Depending on the client contract, reimbursements are sometimes billed to Menlo and sometimes to the client. Then she cleans up some other expense reporting and takes care of a check that's come in from a client. She's still not feeling great, but she's hanging in there and getting things done.

✦ ✦ ✦ ✦ ✦

Jane and Michelle, another PM, are working on the schedule for the upcoming week. The schedule is built in an Excel file that's displayed on a screen at the front of the office. At Menlo, where physical documents—paper and pencil stuff—are still very much in use, the weekly schedule is one of the few things that is digital.

The schedule shows who will be paired with whom, which project or projects they will be working on for the week, and where they will be sitting. Some worktables may be moved around for the week to accommodate changes to the project teams, and pairs may move from table to table on an ad hoc basis if they need to work with other pairs. Computers, and the associated project files and applications, will stay in place for as long as the projects remain active.

Putting the weekly schedule together can sometimes be pretty challenging. For example, when several projects see a need for more resources at the same time, or when a new project kicks in, it can be difficult to make the necessary adjustments. But, based on the steady workload demand for the coming week, this time the number of developers for each project can remain the same.

One thing though. "We should keep the same pairs on Quail," Michelle says, given the "alleged emergency" from earlier in the week. "It might be good to let things settle before making any switches." Jane agrees.

11:30 AM
JARON VOGEL & GRANT CARLISLE
Developers

Rather unexpectedly, Friday morning has turned out to be something of a test for Grant, as almost all of the functionality that he and Jaron have implemented that morning has required several different tests each—tests that Grant is relatively inexperienced at writing. Of course, Jaron is also well aware that their morning seems to be relatively test-heavy, and bearing in mind the feedback that he, Andy, and several other Menlonians have given Grant about his testing skills, Jaron is encouraging him to take the lead.

> Our business in this world is not to succeed, but to continue to fail in good spirits.
>
> —Robert Louis Stevenson

A poster on the wall at Menlo Innovations

Jaron and Grant are still working on Night Sky's sort function, and they need to do quite a bit of testing to ensure that the sorted data is being displayed properly. They read over the lines of code and review the mockups that the client's project manager has provided.

Grant waves at the keyboard. "I'll let you drive," he says, with a chuckle.

PAIR PROGRAMMING

In pair programming, every pair has a driver and a navigator. The driver uses the keyboard and mouse to actually write the code; the navigator's job is to keep a bird's eye view on the task at hand, making sure that the driver doesn't get lost in the nitty-gritty details. In some shops, developers switch roles but typically keep the same role for a specified period of time, which can be an entire day. At Menlo, developers "ping pong" between the roles more freely, as they move from one task to another.

"I actually want you to do this," Jaron says, his tone a bit more business-like than usual. "Oh, because it uses the partial mock?" Grant asks. "Yeah, it will give you a chance to practice it," Jaron replies, as he angles the keyboard towards Grant.

Partial mocking is one of the techniques that Andy and Jaron have flagged as something that Grant should work on. In essence, partial mocking involves testers feeding specific kinds of fake data through a feature that they've finished developing, both to make sure that the feature is doing what they want it to do and also as a way of checking how well their code handles "fringe cases" or weird behavior from a user.

As Grant works on the test, he knows that it's taking him longer than it would take Jaron, even though Jaron shows no sign of frustration. He also knows that Jaron has him doing the test to see what kind of improvement Grant has made in this area since last week. He could undoubtedly go faster if he asked Jaron for help, but he's picked up on how Menlonians conduct themselves at work, so rather than ask, he first tries to do it on his own. He knows that Menlo values curiosity, initiative, and a willingness to learn by doing more than maximizing immediate efficiency.

The next half hour flies by as Jaron and Grant write and pass four different tests for their sorting function, with Grant doing most of the actual programming. As they finish the fourth and final test, some of their fellow Menlonians are heading over to the kitchen to heat up lunch, while others are zipping up coats before pushing past the large glass front door. Grant and Jaron hit save and stand up, Jaron scratching at his chin while Grant lets out a yawn and stretches his arms.

Time for lunch.

11:30 AM
MOLLIE CALLAHAN
High-Tech Anthropologist

Jane had warned Mollie about the Super Secret Red Book. "We should have a process for adding information to it," she had said. "With everybody adding information, everything in there is a jumble."

Jane was right. The Red Book is a mess. The information that's in there is valuable, but with no one responsible for organizing it, much less entering it, finding what you need can take a little effort. Which only means that it takes Mollie a few extra minutes to find the HubSpot log-in data.

When she has what she needs, she logs in and enters the first hot tradeshow lead. Then she opens Word and begins a more detailed lead sheet for that contact. In addition to serving as documentation of the trade show experiment, what Mollie is doing will provide context for others who may interact with the leads.

In her Word doc, Mollie includes her notes from the show, and then with the notes jogging her memory, she adds everything she can recall about her interaction with the prospective client. She Googles the type of software he was interested in developing and adds some of that information to the lead sheet.

If David were in the office, she'd be partnering with him on all of this, but he's out sick, so she'll get his input when he's back in.

11:40 AM
RICH SHERIDAN
CEO and Chief Storyteller

Once back in the office, Rich checks his calendar to see what's on tap for the rest of his day. He has a phone call at 2 pm with a former mentee who's been successful in raising a significant chunk of venture capital to move his entrepreneurial dream forward. Coincidentally, given what Rich just saw at the American Mobility Center, the concept involves using AI to enable a driver to control a vehicle using just their brain rather than their hands and feet.

Then at 3 pm, Rich has a meeting with someone he has never met, but who has been referred to him by a friend of a friend. The guy has recently moved to Ann Arbor for an executive IT position in a local business.

✦ ✦ ✦ ✦ ✦

He checks his emails. There's a request from someone at the University of Michigan BioSocial Methods Center asking if Menlo would provide a student tour. He clicks on the link embedded in the email to see what he can learn about the Center. Looks interesting. Accommodating the group would not only be in keeping with Menlo's commitment to the community, it could also help deepen the relationship Menlo is trying to build with the university's broader Health Sciences domain. He'll talk to Anna about this later.

He also has an email from a big Pharma company. They're asking about a tour too. That's another potentially valuable connection. Then another request from someone else at the University, and an email from the Director of a non-profit medical center looking to connect with Menlo. That one makes Rich smile. The Director used to be a kid in his Sunday school class.

Bob Chapman, the CEO of Barry-Wehmiller, has sent along an announcement of a keynote address he's giving at an upcoming manufacturing conference, along with an invitation to Rich to be a speaker at the event. Bob—the author of *Everybody Matters: The Extraordinary Power of Caring for Your People Like Family*—is a kindred spirit in terms of promoting a people-centric approach to business. Rich forwards the email to the members of the Liberty team: What do you think?

More emails. There's a note from Jane notifying him that enrollment for the health benefits program they talked about just yesterday is now open. As it turns out, Jane is actually working at a table right next to him this week. He jokes with her about sending him an email. (In Menlo's "High Speed Voice Technology" environment, it's generally a no-no to email a colleague when you can just talk to them directly.)

Jane laughs, but then asks, "Will you guys be switching?" She's curious about how many Menlonians will take advantage of the company's just announced decision to pay 100% of the premiums for the HMO option. "That'll be up to Carol," Rich says. Jane smiles and says. "She'll never switch."

Their conversation moves on to Jane's experience trying to navigate the complexities associated with Menlo's health insurance programs. Fifty employees is the line between "small" and "large" organizations with respect to the Affordable Care Act and the Family Medical Leave Act, and Menlo is now right at that threshold. But the legislation and the associated regulations and administrative guidance has created some confusion about who counts or doesn't count as an "employee."

✦ ✦ ✦ ✦ ✦

On Rich's computer, there's a handwritten sticky that reads, "Visit Dave O in Flint," and another that says, "Visit Thomas Z in DC. and NASA!" Next to these notes is a refrigerator magnet that has an ad for Coke, with "JOY!" in large letters.

✦ ✦ ✦ ✦ ✦

Another note taped to Rich Sheridan's computer:

9 Promises for Success and Happiness—Coach John Wooden
1. *I promise to talk health, happiness, and prosperity as often as possible.*
2. *I promise to let all my friends know there is something in them that is special to me and that I value.*
3. *I promise to think only of the best, to work only for the best, and to expect only the best in myself and others.*
4. *I promise to be just as enthusiastic about the success of others as I am about my own success.*
5. *I promise to be so strong that nothing can disturb my peace of mind.*
6. *I promise to forget the mistakes of the past and press on towards greater achievements in the future.*
7. *I promise to wear a cheerful appearance at all times and give every person I meet the gift of a smile.*
8. *I promise to give so much time to improving myself that I have no time to criticize others.*
9. *I promise to be too large for worry, too noble for anger, too strong for fear, and too happy to permit trouble to press in on me.*

11:45 AM
MATT SCHOLAND & ART KLEIN
Quality Advocates

Rick and Kevin have come over to talk with Matt and Art about a card they've been working on. Rick has ripped a sticker in half to produce—the first ever?—half-orange, half-purple dot. He explains that the half/half dot is intended to show that there is work to be done by the client team that won't be completed until Monday. He and Rick figured that in the meantime Matt and Art might want to go ahead and test the items on the card that have already been completed.

11:45 AM
KEVIN F. & RICK COPPERSMITH
Developers

Back at their worktable, Kevin drags the keyboard and mouse over and starts typing out an email to a client. Rick reads along and offers suggestions as the email takes shape. At one point, their conversation briefly turns away from the email to Paid Time Off. Rick says that he has five weeks of accumulated PTO, three of which came from unused paternity leave.

At Menlo, where pretty much everything—including the company's financials and even everyone's compensation—is readily available information, people seem to be pretty open about sharing personal information. In a culture built on trust, caring, and respect for the individual, there's not much need for secrets.

11:45 AM
RICH SHERIDAN
CEO and Chief Storyteller

Finished with his emails, Rich goes off to track down James. As he walks past Anna, he stops to say, "Hey, Anna. About the UM tour request. Very interesting. I think we're starting to see some real connections with these different university groups. Keep me posted." Then he adds, "What can you tell me about that tour request from the Pharma company?"

"It's about their digital team." Anna replies. "They're going from 50 to 100 people in short order, and they're trying to set up as Agile."

"How did they hear about us?"

"Steve Denning," Anna says. Denning is a well-known thought leader in the area of Agile computing. A couple of years ago, Rich was one of the original members of a consortium Steve put together where IT leaders who were implementing Agile practices visited each other's locations to get an on-the-ground view of what different implementations looked like. Based in part on some of the learnings from the consortium visits, Steve went on to write *The Age of Agile: How Smart Companies Are Transforming the Way Work Gets Done.*

"They're still in the process of setting up space," Anna continues. "But they visit one Agile site per month, and they'd like to spend half a day with us."

"Sounds like another good connection," Rich says. His relationship with Steve Denning continues to be valuable in many ways.

12:00 PM
MOLLIE CALLAHAN
High-Tech Anthropologist

A delivery person walks through the office carrying a stack of pizza boxes, and Mollie gets up to follow them to the kitchen and cafeteria area. She and a few other Menlonians set out the pizza, salads, drinks, and eating utensils, as more people converge.

As they wait in line for food, people talk about anything and everything. Some of the talk, although not that much, is work-related, but much of it is just the casual chitchat of friendly colleagues—sports, movies, the news, plans for the weekend.

Some people say that you can get a pretty accurate sense of an organization from a few minutes watching this kind of informal gathering. If that's true, Menlo gives off a good vibe.

12:00 PM
GUINEVERE PROJECT TEAM: ANGELINA FAHS, GIOVANNI STURLA, HELEN HAGOS, KYLER WILKINS & LISA AND JOSIAH H., AND CLIENT
Show & Tell

The Guinevere team has brought lunch into the conference room. Everyone is gathered around the project timeline Lisa has laid out on tables around the room, with its folded, colored sheets outlining each iteration.

As he looks over the display, Carl says, in a playful tone, "Whoa! Too many neat ideas! Too much potential to really screw up project scope and cost!"

While he's eating, Carl continues to run through the tasks and time estimations that define the timeline, resource utilization, and scope of work.

As different stages of the project are considered, the congenial back-and-forth between the Menlo team and Carl continues. "Oh, that's a good point," is heard several times, coming from and aimed at both Carl and the team members. At one point, Carl brings up a three-point feedback process he and his team use in their programming approach. One of the Menlonians who is relatively new to the project says, "Oh, we do that here as well," to which Carl responds, "Great, so we're on the same page." Lisa, who has worked on Guinevere since the start, says, "Yes, when you mentioned it early on, Carl, we went with it."

12:00 PM
BRITTANY MORTON & RACHEL CLEVELAND
High-Tech Anthropologists

During lunch, Brittany and Rachel sit with Anna. Their conversation moves easily across topics. Brittany and Rachel have worked alone for a big chunk of their careers, so they compare experiences. Rachel talks about feeling the need to belong to something bigger than herself, and how Menlo is providing just what she was looking for. "It's exhausting to be engaged with another person all the time," she says, "but it is not a bad thing," adding that, "It's hard to always remember to think out loud!"

Brittany agrees that she too found pairing exhausting yet deeply satisfying. "It's a muscle that you build, to be engaged and work for so many hours with someone else. But now the idea of having any other job feels so lonely, especially when you get stuck!"

Rachel agrees. "You're so much less likely to get frustrated with the work when there's always someone to bounce ideas off of. And working in pairs keeps you from running into trouble as often, because having two sets of eyes gives you constant feedback.

"Partnering also helps keep things from getting too stressful," she continues. "During my first week, my pair partner and I were working on a time-sensitive project and I was starting to get stressed out about it. So, my partner was like 'okay, we need to step away for a minute, let's go for a walk!' and we did. The whole environment here is like that, just positive and supportive."

The conversation drifts to the topic of babies. Josiah's quiet presence throughout the day has been palpable. Neither Brittany nor Rachel have children yet, but they agree that Menlo's practice of bringing babies to work says something important about the organization. "It allows people to bring their whole selves to work, which is good for the organization as well as the individuals," Brittany says. "It shows how much work-life balance is valued at Menlo—a lot more than any other place I've worked before."

Anna, whose daughter had been a Menlo baby, adds, "One of the main reasons why I stayed at Menlo after I had a baby was because I had the chance to bring her to work with me every day. That allowed me to come back to work sooner, which was important to me." Like many mothers, Anna had experienced postpartum depression, which for her was largely alleviated by being re-inserted into a professional environment, while still being able to breast-feed and care for her baby. (One of the few enclosed spaces at Menlo is the "baby room," where Menlo moms and Menlo dads can feed or change their infants.)

Anna adds that she didn't think having a baby on site had had much effect on her productivity, other than the fact that she couldn't bill for the hour or so a day she spent on baby care. The decision to stop bringing her daughter into the office had been hers alone. "She was just too happy," Anna said with a smile. "She got really loud, and I decided to put her in daycare, but it wasn't as if anybody else seemed to mind her happy noise."

Digging a little deeper into what they think and feel about the Menlo culture, Anna, Brittany, and Rachel talk about the opportunities to continuously learn and grow in different roles, about being pushed out of their comfort zone but never feeling like they were drowning on their own. They note that at other organizations making mistakes is definitely a negative, while at Menlo the idea of "make mistakes faster" is seen as a positive part of learning. They also are enthusiastic about the four weeks of paid vacation a year, plus ten additional holidays, and the policy of not working on weekends.

Brittany jokingly sums it up with: "I would only ever leave Menlo if my husband was transferred somewhere else for work."

✦ ✦ ✦ ✦ ✦

Helen Hagos, a developer on the Guinevere project, wrote a blog for the Menlo website on "Building a Culture that Keeps Women in STEM," where she took up many of the refrains that are part of Anna, Brittany, and Rachel's conversation. Helen drew from a *Harvard Business Review* article that identified five biases in typical corporations that pushed women out of STEM. Those biases were:

- the need for women to constantly prove themselves ("every day your credibility starts at 0");
- the need to walk a tightrope of being not too feminine for fear of being seen as incompetent, but not too masculine for fear of being seen as aggressive;
- the "maternal wall" when having a child or children begets questions about commitment and the appropriateness of growth opportunities;

- the "tug of war" in which senior women take the position that since they had to "toughen up" to survive, younger women should do so as well; and
- the self-isolation that comes about as women feel they can't socialize at work for fear of being seen as unprofessional.

None of these, wrote Helen, were experienced by women at Menlo. Why? Not because Menlo offered diversity training or inclusion policies, as helpful and necessary these might be in some other organizations. No, women didn't experience these biases at Menlo because the practices and culture of the organization promote true teamwork and enable everyone's unique talents, interests, and experiences to be brought into play.

Over time, being paired with nearly every other person in your role at Menlo, and collaborating so closely with people in other roles, means that people come to appreciate your skills and talents, so you don't have to prove yourself over and over. As for forcing people into gender-based categories, that doesn't happen at Menlo, because the culture truly does value each person's unique talents, interests, and experiences for how those help the organization succeed. Of course, the "Menlo baby experiment" helps massively to reduce the concern about babies inhibiting one's sense of commitment. And regardless of the tough and toughening experiences women might have to go through in other organizations, at Menlo the collective commitment to treating everyone with respect and dignity makes being "tough" unnecessary.

As Helen concluded: "At Menlo, we want everyone to be balanced, fully-realized human beings." As Anna, Rachel, and Brittany confirm, that goal is enabled by the Menlo culture—by what is said and done, encouraged and supported, day in and day out.

A Menlo dad with baby in an impromptu huddle

12:00 PM
MATT SCHOLAND & ART KLEIN
Quality Advocates

Over lunch, Matt and Art also get into a wide-ranging discussion that starts with the question of whether, and/or how, introverts can survive in a place like Menlo, with its emphasis on collaboration, exemplified by practices such as pairing and Daily Stand Up. The question comes up often at Menlo, where, according to James, there are only three extroverts on the team of 50. (He might be kidding about the exact number, but not about his basic point.)

Matt sees no problems, so long as you have the right understanding of introvert. "An introvert doesn't need a large group to get themselves energized. They recharge by being by themselves or with a small number of people, but that doesn't mean they're anti-social. At Menlo, there are a lot of people around, and sometimes we all come together, like at Daily Stand Up, but mostly we interact with relatively few people, and it's mostly focused on work. We mostly talk about something defined and work-focused."

Art: "Menlo isn't for everyone, just as Menlo isn't right for every client. But I'm not sure we want to get hung up on this introvert-extrovert thing. Who likes to be obligated to being identified as one thing or another? I've been reading about how we get into ruts of identifying ourselves, or letting ourselves be identified, as being just this one type, in terms of personality and how we react to things. But it doesn't have to be that way. So, sure, sometimes I'm an introvert… but maybe my preferences are changing over time."

Matt turns philosophical. "I think you have to look at things more broadly. You have to think about the whole person. What we have here at Menlo is a system founded on ideas like what is the nature of goodness and how human beings need to relate to one another and how do we expand our sphere of moral consideration and what is a healthy community. Sure, these are pretty high-level ideas, but at a very practical level, Menlo represents a really great experiment in what is a healthy workplace."

> **MAGNANIMITY**
>
> In high-performing organizational cultures, magnanimity runs deep. It's about "we," not "me." It's about truly valuing the different gifts and talents that others bring to our collective endeavors.

Art brings up Barry-Wehmiller, Bob Chapman's company. B-W is known for taking rust belt companies slated for bankruptcy, and instead of shutting them down, firing the employees, and selling off the assets, turning them around. Like Menlo, B-W's whole approach to business focuses on the people and on treating employees with respect and dignity. They've been very successful at getting that mindset in place and then making those broken companies profitable again—with the same people.

Matt: "It makes sense. The employees really appreciate B-W's investing in their factory, and they show it by their loyalty and their level of responsibility. At Menlo, we all take responsibility for our success and our clients' success. We're all on the same page, and we're really careful to bring in people who fit in with that. It really is such a cool experiment. We can see how this approach affects us and our clients, as human beings. I know this sounds a little grand, considering how small a company we are, but this kind of model for how people can work together has implications for society as a whole."

Art: "The way we work together helps us all be calmer. We actually calm each other down. This has a broader effect. It's one of the reasons other companies get excited about Menlo. I compare Menlo to a small organic farm where I once worked. We do our work in community. We're not working away alone in our own cubicles. We're connected. We don't have superheroes. What matters is taking something on together."

Art and Matt mused on the potential impact of the growth that James and Rich have projected. "I think we can go to 150 people with minimal changes," Matt says, "although, with that many people, we would only pair with the same person every couple of years so that would not be great. But three groups of 50 is not the same. Even bigger could be done and has been done in other companies, but that would require more significant changes."

Art: "I go back to that organic farm I worked at. If you want to have a giant farm to feed everybody, it's hard for it not to be optimized for efficiency. And that's when you start to squeeze the human person out of your thinking."

The conversation shifts again, to the most important characteristics of people they each want to work with. Matt says, "Courageous people—people willing to have uncomfortable conversations with clients, with me, with co-workers. Number 2. Process-oriented people. People who want to make the best process for adding value, not just the best product. Number 3. Kind people. Kindness is crucial."

Art: "My Number 1. Great storyteller—I really appreciate the way that the Devs and PMs present what we do at Show & Tell. They incorporate the important points into a story. This is what we did this week, this is why it matters. They connect to the characters involved and get the clients connected. They create empathy. Number 2. Values. Doing things for some greater good, caring about more than their own personal needs. Number 3. Someone interested in how human organizations can be healthy and vibrant."

The question arises, Does Menlo's commitment to hiring people who will fit well with the culture limit cultural diversity? Matt's sense is that hiring for kindergarten skills doesn't prevent Menlo from having a diverse workforce. "We don't keep strict data on this, but based on our Levels board, it seems to me that women do very well at Menlo; women are a significant majority of the Senior Consultant and Principal levels. I do know that we are recruiting from a wide range of sources and recruiting continuously. The main filter for the Extreme Interview is just expressing interest in Menlo. Instead of a traditional interview where we talk about experience and work background, we bring in a large number of candidates and observe the candidates pairing with one another. This gives us insight into how our candidates interact with each other. I think this is a much better way to evaluate candidates and gives all our candidates a fair opportunity to demonstrate what they offer. I think the structure of our onboarding process gives us a diverse team, which you can prove just by chatting with a few of our teammates. We have folks from a wide range of backgrounds, economic statuses, and educational experiences."

12:00 PM
KEVIN F. & RICK COPPERSMITH
Developers

Kevin and Rick grab some pizza and find a table in the cafeteria area. Maybe it's something in the air, but they too start talking about what is apparently a current hot topic —the stereotype of software developers as introverts.

Rick says, "You're an extrovert. You were excited about having friends come over to your place last night." "So much for all developers being introverts," Kevin jokingly replies.

Kevin tells about the time he was assigned to give a tour of Menlo (with a partner, of course). "I'd never even been on a tour myself. And then there it was, just on the schedule one day: 'Kevin, you're going to be giving a tour on Friday.' I had no idea I'd ever be asked to pair on a tour. We don't exactly have the most official onboarding process."

"It's much more official now than when I started," Rick replies. "I wasn't even told that I succeeded with my three-week interview process. I had to ask! On the last Friday I was like, 'So I noticed I'm on the schedule for next week, does that mean I passed my three-week?'"

Rick reminisces about the floods that Menlo periodically experiences, one of the less than ideal results of being in a basement. "One time, when some of the overhead pipes were leaking, the landlord put up some big tarps to catch the water until he could get the problem fixed! Talk about a band-aid solution! But I tell you what—we all jumped in and we saved every computer in the place."

What's important to an organization's culture is often conveyed by the stories its people tell and re-tell. What are you hearing in these stories?

12:30 PM
RICH SHERIDAN
CEO and Chief Storyteller

Having spotted James working with the Guinevere team and Carl, the client PM, in one of the conference rooms, Rich headed back toward the kitchen, chatted with a few folks, put pizza and salad on a plate, and found a place to sit.

And then, there's James, coming out of the conference room. Rich waves at him, and James heads over. Rich has been looking for him to discuss an apparent problem with one of their current projects. The Menlo folks are continuing to plug away at it, but as Rich has been told, there's a disconnect at the client end.

"How did the last Show & Tell go?" Rich asks. James sighs. "Not great. Our team raised some issues, but one of the key client guys said, and I quote, 'I don't do issues in real time.' Then he asked for a break." With a sigh of his own, Rich says, "Great."

Pat, the Menlo PM on the project, comes into the area carrying his lunch. Rich waves him over. The discussion continues.

Rich: "What about us? Are we performing?"

James nodded yes. "From their point of view, we're better than anybody they've worked with before. But from our perspective, we're having trouble getting the product where it should be. But it's their platform that's poor and we're trying to fix or work around what they've already brought."

The discussion continues. Pat and James lay out a litany of issues on the client end. "They don't have a common vision." "Bill's a micromanager." "They're not giving us all the information we need." "Bill doesn't trust us." "Bill doesn't trust anybody but Herb. Herb has all the facts. When Herb doesn't like the direction, the facts change." "Herb says that Charles is an idiot. Charles is new on their team, but he's trying to figure things out."

Rich: "How do we get all the stakeholders together to figure out what everybody needs? I don't see how we get past where we are now without seriously butting heads. If this thing blows up, it will be on us, so maybe we need to have a serious conversation with Bill. Maybe this relationship just doesn't work."

James, usually the pessimist, says, "Every week, I think the relationship is getting better. We can continue to learn, to build trust."

Rich, the eternal optimist, replies, "I think the relationship is getting worse. What option do we have besides putting the problems in front of Bill?" Wistfully, he adds, "I care about Bill. And somebody I care about has a big problem and he's not facing it."

James and Pat aren't ready to go into full-confrontation mode yet. They explore different resourcing arrangements. Could someone else from the Menlo team join the project and have a positive impact? Would it be fair to put anyone into such a position? Did they want to send someone to a new, difficult client simply because the client did not have faith in the other, equally competent, Menlonians?

The conversation keeps coming back to problems on the client end. "They're questioning my professional competence," James says. "They're challenging our process. They don't want to have anything to do with what Menlo does!"

Pat comments that there's more at stake than bad software. This client seems to rely on fear-based management, and there's a danger in allowing that to be introduced into Menlo's process.

Back and forth. Rich makes the obvious point: "Just to be clear. I don't want to lose this project." James and Pat agree, but would they rather lose a client than damage their reputation by delivering bad software? Are those the only alternatives? Is there a better option?

At that point, Michael, a young developer, walks up and hesitatingly asks, "Can I interrupt for just a second. The team decided to hire Vaughn. We made him an offer today."

"Great!" James says.

Michael demurs a bit. "He's a bit quiet, but we think he can learn to be comfortable speaking up."

James: "You think he's quiet? I find him to be not too introverted. He says hello in the morning and goodbye at night. He makes eye contact. He walks right up to you."

"Which level would he start at?" Rich asks.

"We are thinking A1 (Associate 1—the entry level position)."

"Good job," they congratulate Michael, who walks away. His news was a welcome break, but the conversation immediately turns back to the issue at hand.

"What swirls around in my head," says Rich, "are the possibilities. I see this primordial soup of who they are and who they know and who we are and who we know and all the possible connections solidifying if we get this right."

"And then you see Bill throwing Menlo under the bus if we get this wrong," says James, again in pessimist mode. "And, for the record, let me tell you where I'm at. You two don't want to leave me in a room alone with Bill right now. He's been goading me."

Pat chimes in, "Don't tell me about goading. My anxiety level goes up whenever I see him. I don't want to be here if he's been around. I'm really afraid if he says something to me, I'll say something snarky."

The threesome bandies around more options, and finally, Rich leans back in his chair. "I don't see a solution yet, so it's probably best if we don't act yet."

James turns toward Pat, and says with a touch of sarcasm, "He's just trying to cheer us up."

1:00 PM
JARON VOGEL & GRANT CARLISLE
Developers

Jaron wanders back over to his worktable. Grant is still in the kitchen, filling his cup with coffee. Jaron glances at the storycard he and Grant have been working on. As Grant walks over to sit back down, Jaron smiles and holds up the card. "Are we done?" Grant asks.

"I think so," Jaron replies, with more than a hint of excitement in his voice.

"All right!" Grant answers back. The two exchange a high-five and head over to the Work Authorization Board. They take one of the orange dot stickers from the wall, stick it on the storycard, and pin it in the Night Sky swimlane. It's ready to be reviewed by another pair of developers.

As they scan the rest of the cards in their lane, they realize that they seem be stuck. All of the remaining cards for the week require information to be passed from one page of the application to another. Unfortunately, this feature isn't finished. It is in fact what Al and Travis, the other pair of Night Sky developers, have been working on all day.

Looking at the storycards in Al and Travis' lane, Jaron points to one at the top of the lane, a card that involves stylizing the headers of some columns of data generated by the sorting function that they've been working on. "What do you think about working on this one?" Jaron asks.

"Let's see…" Grant says, as he takes a closer look at the card Jaron is pointing at. "Yeah, let's do that one!" Jaron pulls the card off the board and he and Grant walk over to Al and Travis, who are locked in an intense staring match with their computer screen. The computer seems to be winning. Jaron says, "Hey guys!" Turning away from their screen, Al and Travis glance at the card that Jaron is holding.

"Grant and I think we should work on the style card that's in your lane" Jaron says. "We think that adds more value at this stage" Grant adds.

Al looks over the board before turning back around to look at Travis, who nods his head in approval. "Yeah, that's a good call," Al says. "Go for it."

With a nod and a smile, Grant and Jaron walk off to work on their new storycard.

1:00 PM
MOLLIE CALLAHAN
High-Tech Anthropologist

On Mollie's way back to her desk, she and Lisa talk about the hot leads Mollie has identified from the trade show, and how to follow up with the rest of the Liberty team. Anna, also on her way back to her table, joins the conversation.

"What I'm concerned about is our workload. We're already stretched thin," Mollie says.

"I think we should push and set up meetings even if we don't have the resources. That's the point of Liberty," Lisa, the Guinevere PM, responds. It's a hard call. On the one hand, Menlo practice is to make sure people have the time and resources before making a commitment. On the other hand, the Liberty Project is all about generating new business that will lead to more revenue coming in that would allow for more resources, and by extension, more time.

"Okay, I'll go ahead and get those emails out," Mollie says.

✦ ✦ ✦ ✦ ✦

Back at her table, Mollie finishes typing up her notes from the show, then turns her attention to contemplating the emails she might draft to each of the leads. She's starting to feel the day catching up with her, so she decides to go on her afternoon "Walkie" and grab a cup of coffee.

Getting up from the table, she walks over to get her coat and heads out of the building. The crisp fresh air feels good as she walks quickly to a nearby coffee shop—a favorite of Menlonians—where she picks up a latte and returns to the office. The whole excursion trip takes just 10 minutes, but it's enough of a break for Mollie to feel re-energized.

Once more at her worktable, she settles in to continue following up with the leads. She reads through her notes one more time and decides that she's comfortable enough with her understanding of their needs that she's ready to draft the emails.

1:00 PM
BRITTANY MORTON & RACHEL CLEVELAND
High-Tech Anthropologists

By the entrance to the office, the Cherubim team has set up some movable pin boards to create an enclosed space for the Show & Tell; the boards are covered with posters showing design mockups. Inside the temporary meeting space, the Menlo and client teams stand around a couple of tables and make small talk. Included today from the Menlo side are Brittany and Rachel, both HTAs; Michelle, the PM; and two developers. Three people from the client side are expected, but at 1:00 only two have arrived.

The client in this case is a local religious organization whose goal is to build connections in the community; the project involves building a platform to help manage events. At 1:15 the third member of the client team arrives, and the meeting gets started.

First up are the Menlo developers, who show what they've done so far and answer questions about procedures they are implementing. At 1:45, Rachel and Brittany begin their presentation, with Rachel taking the lead. At one point, there's considerable pushback from the clients on one of the design mockups and the features it includes. The discussion reveals a difference between the conclusions the HTAs have drawn from their observation of the end user and the client's view of how the software will be used. It isn't unusual for this to happen. Indeed, the reason Menlo uses HTAs is because clients—who may not be end users themselves—often have a mistaken view of how the actual end users will ultimately work with the new software.

As the back and forth unfolds, Rachel and Brittany are clear about what they learned from their interactions with the end user—the primary persona—but they are never defensive, and the discussion on both sides is thoughtful and respectful. Everyone in the meeting is clearly trying to make sure that the product Menlo delivers meets the client's needs.

Finally, Michelle, the PM, says, "Why don't we run another experiment to determine what the primary persona would gain more joy from?" She suggests using Menlo's network of clients to quickly sample some end users who use event management software on tasks very similar to those Cherubim needs to carry out. The client team agrees and just like that, the disagreement is settled. Focus on the end user. Get data quickly. Experiment.

1:10 PM
GUINEVERE PROJECT TEAM:
ANGELINA FAHS, GIOVANNI STURLA, HELEN HAGOS, KYLER WILKINS & LISA AND JOSIAH H., AND CLIENT
Show & Tell

> *Build projects around motivated individuals.*
> *Give them the environment and support they need,*
> *And trust them to get the job done.*
> -- A Principle behind the Agile Manifesto

At 1:10, the online link to the client team is shut down, but Carl and the Menlo team continue working through the Guinevere long-range plan. The work goes on until 3:00, when Carl finally leaves.

Before the team breaks up, Helen updates the storycards for the next iteration, based on what occurred during the Show & Tell. Lisa lays out the pair assignments and estimated hours for the next week.

A most satisfying Show & Tell.

1:15PM
JANE COLLINS
Project Manager

Back at her worktable after lunch, Jane has a few more things to do before she leaves for the day. She verifies an invoice amount with a client, and reviews some of the ACA terms that will soon impact Menlo. Michael and Vaughn come by to tell her that Vaughn is accepting his offer! Jane congratulates him and prints off a spreadsheet that outlines Menlo's different levels and pay grades, so they can discuss where Vaughn would start.

Michael asks what steps they need to take to get Vaughn started as soon as possible. Normally, Jane would get the necessary W-4 and I-9 forms ready immediately, but she's definitely not feeling well, so she says they can take care of everything first thing Monday morning. On Monday, Vaughn will officially become a Menlonian!

1:18 PM
KEVIN F. & RICK COPPERSMITH
Developers

Kevin and Rick take a bathroom break at the same time, leading Rick to comment, "That level of synchronization is unusual."

Pair partners work so closely together that if one of them takes a break, the other is almost entirely unable to continue working. This discourages too many soda runs, or too much time spent on the phone, which can leave your partner twiddling their thumbs. Bathroom breaks are inevitable, but accidental "synchronization" means maximum pair partner efficiency!

✦ ✦ ✦ ✦ ✦

Break over. Kevin and Rick discuss an architectural change to the code, debating how large or small that change should be. What problems with the code are they trying to solve? Which of those problems has the highest priority? How can they avoid redundancy? Whichever way they go, they'll need to write tests for the changed code. Depending on how large a change they make, the time involved will range from four to eight hours.

"This makes me want to strategize it," Kevin says. "Do we have time to strategize?" Rick asks.

"There's a potential for it to spiral if we don't handle it right," Kevin replies.

"Okay," Rick concedes. "Do we need a whiteboard?"

1:15 PM
RICH SHERIDAN
CEO and Chief Storyteller

After lunch, Rich walks over to Asha, an HTA, and a whiz with PowerPoint. Rich needs her help for a presentation he has to give on Saturday.

Settling in beside her, he says, "Let me give you the background."

Rich's second book, *Chief Joy Officer,* will be released soon, and he's reasonably sure that his 40 talks per year will increase when it comes out. That certainly happened with his first book, *Joy, Inc.* With that in mind, he's upping his game by working with a professional public speaking coach based in Detroit. (The coach is himself a successful public speaker, delivering 160 talks per year. Rich knows of one day when he spoke at a breakfast in Detroit, took a private jet to Montreal for a talk at lunch, then flew to Rhode Island for a talk before topping the day off by helicoptering to New York City for an evening talk.)

On Saturday, Rich is going to film a three-minute video at the coach's Detroit studio. The video will be uploaded to a Rich Sheridan website so that people potentially interested in booking Rich can get a flavor for what kind of speaker he is.

Moving on, Rich says, "You're way better at PowerPoint than I am, but let me show you where I'm at so far." With that, he plugs a data stick into Asha's computer and opens the PowerPoint deck he has already produced. There's a slide showing a timeline of Rich's working history that graphically displays his rise up the corporate ladder, then the fall-off in what he describes as the "trough of disillusionment," and the period of time when he was searching for a way back to fun at work. There are slides depicting some of the books and thinkers he drew from in developing the ideas that ultimately led from the Java Factory to Menlo—books by Tom Peters, Peter Senge, Peter Drucker, and others.

When he gets to a slide showing an airplane, he explains, in an increasingly excited tone, how he'll use that image to talk about the workings of a high-performing culture.

"I'm a pilot. And I was thinking about how flight is a function of four countervailing forces. There's the weight of the plane. And then there's lift. There's drag, and there's thrust. For a plane to fly, lift has to be greater than the weight, and thrust has to be greater than the drag. An organization is like a plane. The lift is human energy. For the organization to take off, that lift—that energy—has to be greater than the weight of bureaucracy. Thrust is purpose. In a high-performing organization, the thrust comes from a worthy, externally oriented focus on the customers, clients, and end users the organization serves. For the plane—the organization—to fly that thrust has to be greater than the drag created by fear."

Asha seems enthusiastic about Rich's analogy, but she questions the static plane image that Rich is currently using. "Shouldn't it look like the plane is taking off? Shouldn't there be a runway?"

"Yes!" Rich replies with great energy. "When you're rolling down the runway, and you pull back on the yoke—just a little bit—and the plane takes off—every time I do it, there's no greater thrill! And what's important is that if the plane is designed right, with the right lift-to-weight ratio and the right thrust-to-drag ratio, then all it takes is this small input of pulling back ever so gently on the yoke and you're flying. The same with organizations. If they're designed right, with the right ratios, then it takes just a little leadership nudge to get the whole thing flying.

"You know," he adds, "When Orville Wright was asked about human flight, after Kitty Hawk, he said, 'You can't build an airplane that will ever carry more than two people.' He had flown, but he hadn't discovered or articulated all the relevant principles behind flight. We are right now discovering the relevant principles for human organizations. We don't have them all yet. But when we discover them and apply them, we will fly to heights and distances previously unimaginable."

1:17 PM
MOLLIE CALLAHAN
High-Tech Anthropologist

Mollie opens the follow-up email to the first hot lead. She's aiming to make a personal connection, so striking the right tone is crucial. She wants to convey the sense that her email is coming from a person, and an organization, that really does care about the person, and the organization, on the receiving end.

As she types away, Mollie knows that she's probably putting in more detail than they will include in the final version, but later it will be easier to take information out than to add it in. She works and reworks her draft until she feels as if it could potentially be sent out, but now it's time to pass it on to the other Liberty team members for their reactions.

1:18 PM
MATT SCHOLAND & ART KLEIN
Quality Advocates

On Glassdoor, most of the evaluations of Menlo are positive, but in the past the few negative ones seemed to come mostly from people who worked for Menlo as contractors. Basically, they didn't like being contractors and would have preferred to be employees. Not directly in response to the Glassdoor comments, but certainly in response to a sense that they might lose some good people who were working for them, Menlo had made a switch. So, today, there are no contractors at Menlo. There are "full-time," "part-time," and "variable" employees.

Talking about the switch, Art says, "I wasn't at Menlo before the switch. But my guess is that a lot of the Glassdoor comments arose because of a framing issue. These people were asking if they were really part of the team or could they be fired on the spot. That's a lot of uncertainty to have to live with."

Matt, who had started out as a contractor, agreed, in part. "I do think it might have been an issue for some people whether they felt part of the team or not. But when I was a contractor, I definitely felt part of the team. For me, it wasn't a fear of being fired so much as just never being sure if there would be enough work to keep me on payroll. There were a couple of times when I was asked to stay home because of a lack of projects.

"But there were upsides to being a contractor. The hourly pay was better. You didn't get health benefits, but if you had them through your spouse, you didn't need them here. And the flexibility was important. On the other hand, it was a pain from a tax point of view. You had to be your own LLC. So, it was mixed."

Art says, "I think the recent pay raise will lead to more positive comments on Glassdoor. The raise was very well received. And better compensation should help with recruiting. We've got a lot of clients who want more time from us, so we need to add people—the right kind of people. That's a good problem to have."

Then Art adds, "But for me, frankly, this is a dream job. There's really nothing structural that I dislike about it."

✦ ✦ ✦ ✦ ✦

Matt gets called over by another pair, but he's back quickly, and he and Art start work on a test of the auto-matching code. They run through a number of functions—edit, delete, resolve, etc.—to make sure everything is working properly. They decide that they need more tests before they can say that the auto match is done.

They decide to run a selection of test cases. Immediately they are immersed in an intense, focused back and forth about data, results, numbers, and then about the "anticipation discount." They check to see if the invoices are showing up in the correct buckets. The client has been doing some of this manually, and when Menlo delivers the final system, the process will be fully automated.

1:47 PM
KEVIN F. & RICK COPPERSMITH
Developers

Kevin and Rick have strategized their approach to the architecture change, and now they've started the work. They want to get the necessary change done in less than 12 hours, and the way to do that is to "make mistakes faster." The objective is to detect errors in the code sooner rather than later. That's how you keep from falling down the rabbit hole.

Rick passes the keyboard to Kevin. There's a natural flow to their work. Isn't this the way software development *should* work?

2:10 PM
JARON VOGEL & GRANT CARLISLE
Developers

> *Welcome changing requirements, even late in development. Agile processes harness change for the customer's competitive advantage.*
> -- A Principle behind the Agile Manifesto

Jaron and Grant have been hard at work on their new storycard. By 1:30 they had finished stylizing the headers on the app, and clean, square, navy boxes now define each section header at the top of the application.

At 2:00 they were supposed to have a phone conference with the client PM for Night Sky to update him on the progress they had made over the last couple days. But when they try to call the project manager, the mic connected to their computer isn't working. Jaron quickly fetches another mic from the storeroom, but this one doesn't work either!

Gremlins have apparently infiltrated Menlo, so a different tactic is called for. They walk over to Al and Travis: "We can't seem to get the mic to connect to our box properly" Jaron says. "Would you be willing to let us steal your computer for a few minutes?"

"Yeah, that's fine," Travis says. "We need to rethink how we're approaching this problem, anyway."

So now, the Skype call to the client goes through. Grant and Jaron apologize for the delay, and then they share their screens with the PM, guiding him from one page of the application to another, showing off quite a bit more functionality than at their last demo a week ago. The PM is pleased with the progress, but then he says that the general presentation of the information on the screen still looks "cluttered" to him. This despite the fact that the layout of the application is identical to the layout detailed in the mockups that the client provided to Menlo earlier.

In particular, the PM says that he's just not sure that the headers look good in their current form. Grant comes back with: "We can put the headers at the top of the page like you mentioned, but I think that you'll end up not liking it. It will still make the page too cluttered." The PM goes silent, before finally conceding, "Yeah, I think you might be right." The discussion moves on to how the team should proceed.

Jaron has been mostly quiet throughout the discussion. Reflecting back on the situation later, he decides that while Grant certainly did nothing inappropriate, he (Jaron) probably wouldn't have handled the situation exactly the same way. While sometimes it's your job as a developer to push clients in a certain direction, it certainly isn't always the case. Part of the problem here is that none of the Menlo HTAs were involved in the initial design process. This has caused some issues before, and it will probably continue to plague this project.

Again, it's not that what Grant did was inappropriate. He was adjusting to changes the client thought he wanted, and in the end, it's all about delivering to the client's expectations.

2:00 PM
JANE COLLINS
Project Manager

By 2:00, Jane has packed up for the day, ready to head home and hopefully start feeling better. She chats briefly with Anna, Rich, James, and Lisa, working at adjacent tables. "Have a great weekend!" she says. "Go home and get some rest," and "Hope you feel better," come back at her.

Jane is tired. Her day may have been shorter than usual, but she's been busy. She's taken care of payroll—which in a very real way is taking care of her colleagues—and she's helped several of them resolve issues that are important to them and to clients.

There's joy in that.

2:00 PM
RICH SHERIDAN
CEO and Chief Storyteller

Rich leaves Asha, who has taken on the task of upgrading his PowerPoint. Looking for a bit more quiet, he sees that one of the conference rooms is now empty. Sitting down at a table, he makes his 2:00 call to Robert Springer, the former UMichigan student he once mentored.

Robert has created some exciting technology in the brain/computer interface space and has raised some start-up funding to get things going. The sad thing, as far as Rich is concerned, is that the Ann Arbor investor community felt the technology was a bit too far in the future for them, so Robert's main investors are from Cambridge, and they want Robert to build his organization in the Cambridge area.

Robert takes Rich's call immediately. He also has some of his staff with him on the call.

"It's been almost eight months since we chatted," Robert says. "What's new in your life, Rich?"

"Two great things," Rich replies. "Our second granddaughter was born. And my second book got finished. It should be in our hands in a week and a half. December 4 is the official launch date. But how about you?"

Robert gives a quick review of what's happening with his start-up process. Things are moving along at a good pace. Exciting things are happening. But Robert is particularly interested in talking to Rich about how to lead in a non-traditional way, a Menlo way. "Obviously, I've read your book, and I'm looking forward to reading the new one, but what else should I be reading? he asks.

"I'd strongly recommend *Leadership and Self-Deception,* by the Arbinger Institute. It's one of those management books told in story form. The main character is a guy called Lou, who's the kind of leader you're talking about. And then, if you want to know how Lou became the kind of leader he is now—and of course this is all fiction—you'll want to read *Anatomy of Peace* to learn about his journey and what brought him to the insights of *Leadership and Self-Deception.*"

Robert: "One of my favorite books is *The Five Dysfunctions of a Team.* You gave it to me when we first met, and it's been transformative for me. It's helping me deal with what's happening in my team right now. One thing I've learned so far in this venture is that having an organization of 12 people isn't the same as an organization of 6 or 7. It's so easy for things to get out of control."

"I really admire Lencioni. Another book of his that I fell in love with is *The Ideal Team Player.* According to Lencioni, the ideal team player is hungry, humble, and smart. And not IQ smart, but people smart."

Robert brings up something he has recently read about the difference between a wartime versus a peacetime CEO. "We have a great team, and a good culture. But we want to become a great culture. I'm not sure if everyone has bought into that."

Rich: "You do what you have to do. Give people time to adjust. But if they don't, you really need to let them go. When we offer our tours here at Menlo, one of the most typical questions we get is, 'How do you build a great culture?' I say, 'You can either have an intentional culture or a default culture.' Default cultures can work for a while. But when they stop working, you don't know why. You don't know the relevant principles. You want an intentional culture.

"Now, there are two elements that support an intentional culture. The first is hiring. Who are you hiring? The second is storytelling. What stories are you telling one another?"

Robert: "When we onboard people, we have a culture deck and I go through it. But I'm not sure we have much storytelling."

Rich doesn't say it, but while a culture deck might be a good first step, if it isn't backed up by what people actually do on a day-to-day basis, and if the culture's foundational concepts aren't captured in the stories people tell, Robert isn't going to get to a "great culture."

Robert turns the discussion to compensation, one of the issues he's been struggling with.

"Compensation is definitely consequential," Rich says, "and if you screw it up, it can really plague you." But the problem, according to Rich, is that traditional management thinking and practice make compensation the "go to" lever for most managers. That's too limited a view.

Rich brings up another book, *Influencer* by the folks at VitalSmarts. "One of the things they say in *Influencer* is 'your world is perfectly organized to create the behavior you are currently experiencing.' We have a poster with that on it hanging in several places around Menlo. Now what do they mean by that? They mean that whatever outcomes you are getting, you've basically designed your system in such a way as to produce those outcomes. If the outcomes are lousy, your system design is lousy.

"Here, this will help you visualize the insights of *Influencer*. Got a piece of paper? Okay. Draw a box that has three columns and two rows. It's just a simple 3x2 matrix. Label the first column Motivation, and the second Ability. Now label the first row Personal, the second row Social, and the third row Structural. So, the upper left box represents Personal Motivation.

"Now, let me give you a simple example of how this model works. When I was in my early 40s, my knees started to hurt. So, I decided I needed to get in shape and strengthen my legs and knees. I was going to work out.

"But what happened? Life happened. One day I was too busy. The next day I had a meeting scheduled. The next day it was something else. I was like almost everybody on December 31 who vows to get healthy in the New Year. Good intentions. Poor follow-through.

"Time goes on. Now I'm in my 50s and I say, 'I'm going to be healthier in my 60s than I was in my 50s.' But the same thing happened. The point is that personal motivation isn't enough.

"Then I ran into this guy, Bud. We'd known each other for almost 10 years, although we didn't know each other that well. Back when we first met, Bud had weighed like 350 pounds, and now he was around 170. So, when I ran into him again and we got to talking, I asked him how he had done it. 'I got myself a personal trainer,' he told me. 'Someone who would stay after me. Made all the difference in the world.'

"Now the personal trainer, in the Influencer model, is Social Motivation and Social Ability. The trainer provides social encouragement that helps you stay with your fitness routine. And the trainer also provides the right fitness routines and so on that help you get fit in the way that's right for you.

"The point is, according to VitalSmarts, unless we address at least four of the six influence boxes, we're not likely to bring about the results we truly want. But what most management teams do is focus on only one of the six boxes. They focus on the one box of Structural Motivation and spend all their energies trying to design the perfect compensation system. They're not asking the right question. The right question would be something like, 'what can we do to also create Personal and Social Motivation?'

"The VitalSmarts folks think that the six boxes are not all the same size, in terms of the influence they have on the outcomes. They think Social Motivation and Social Ability are the most important. That's why here at Menlo we insist on pairing. Working with someone else provides both Social Motivation and Social Ability. Same thing with storytelling. It's social. Now I'm not suggesting that you copy what we do at Menlo—for one thing, you're not a software development company—but you can think about how to build Social Motivation and Social Ability in a way that works best for your organization. For example, is there a way pairing can work for you?"

Robert: "That's hard for us. We're all so specialized in our areas of expertise."

Rich: "Well, maybe you should try some small experiments. Don't try pair working for a year. Try it for four hours! See what you and your team learn from that."

As the conversation pushes toward 3:00, Robert says, "We'd like to visit you and maybe go on a tour. And we'd really like you to visit us. I think you'd be proud to see what we've taken from you."

It was a good conversation.

2:05 PM
ANDREW MUYANJA & ERIC COOK
High-Tech Anthropologists

Andrew, the HTA whose blue sneakers prompted the "competition" with James at this morning's Stand Up, is working with Eric at a table toward the back of the office, near where the Stand Up took place.

At this point, they're on the third round of their project—the Horizons project— a process which includes cleaning up the different features and ensuring the software design is exactly how the client needs the finished software to look. To get to this point, they've gone through multiple interviews with clients, key stakeholders, and potential users. They've also gone through two rounds of design and tested those designs on end users.

Eric has a Ph.D. in Information Sciences, and before coming to Menlo he taught at the University of Michigan. He's been at Menlo for about three-and-a-half years. Andrew had been at Menlo for four years. His degree is in Engineering and Entrepreneurship.

Reflecting on their current project, Eric says, "This may be my favorite project of all those I've worked on." The project brings many different possible end users to the table, so finding a way to meet all of their needs has been a challenge, but also a growth opportunity. Eric and Andrew both feel that they learned something new each time they met with the potential end users. "It was great to do a deep dive into the specialized needs of all the people at their company," Eric adds, "and then to use all that information to design a very coherent system."

Andrew agrees that he's learned a lot from this project, but his favorite Menlo project was a consulting engagement where he taught a business how to practice high-tech anthropology for their clients. It allowed Andrew to take an in-depth look at what he does on a day-to-day basis, understand the strategic value of it, and share it with others. How often does someone get to truly reflect on why their job even exists?

As HTAs, Andrew and Eric's work has varied dramatically based on the types of projects they worked on, the end user needs, the level of client involvement, and the feedback they received from different stakeholders. With all of these moving parts in the HTA process, one of their most significant tasks was to build consensus between these groups of people.

"It's kind of like family therapy," Andrew remarks. HTAs often work with clients to help them understand that their wants and the end users' needs do not always match up, but that the most effective software will be designed with the end users in mind. That conversation is rarely an easy one, and Andrew and Eric agree that one of the more challenging parts of their job is when a client chooses to ignore their findings.

Eric: "You want to do right by the client, you want to do right by the users, but ultimately it's the client's money and their choice. We're just here to be helpful."

Eric also notes that sometimes HTAs have to do what's best for the client, even if it means less business for Menlo. For instance, one of Eric's clients wanted a huge development project done, but after researching the project, Eric and his partner found that software was not the solution to the problems the client was facing. The pair shared their opinion with the client. "The best outcome was we got ourselves fired from the project. It was best for the client."

2:05 PM
MOLLIE CALLAHAN
High-Tech Anthropologist

"Hey Lisa! Do you have time to look at a quick email?"

Lisa walks over and skims the email on Mollie's screen. Referencing the particular need the potential client had discussed at the show, she is not encouraging. "We do get asked about those skills in pre-sales on occasion, but we don't currently have the skills on board to fulfill that particular need."

"An HTA, post-discovery, is in a position to help a client make decisions even if we don't ultimately design the software," Molly replies, suggesting that there still might be an opportunity to work with this company. "Does the email look okay otherwise?"

"Yeah, it looks good." Then, before heading back to her table, Lisa adds, "There might be an opportunity for the team to work together to decide who the best person is to reach out to the other leads."

With the email ready to go, Mollie realizes that, somehow, she doesn't have the contact's email address on her spreadsheet. She pulls his business card from the stack on her table—and no email! Okay! That's what phones are for. She dials the contact's number, chats briefly with someone on the other end (not the contact himself), and comes away with his email address. She BCC's the Liberty team, including David, who actually had the original contact with the lead at the show. Hit send.

With the email gone, Mollie inputs the action taken into HubSpot. The program prompts her to assign follow-up tasks to particular members of the team, but she isn't sure who that should be, so she decides to ask Lisa.

2:30 PM
ANDREW MUYANJA & ERIC COOK
High-Tech Anthropologists

Eric and Andrew are down in the weeds of the Horizons software design, discussing things like text box size and layout uniformity. They want to make sure everything is perfect before the project moves to the development phase.

"What else should we be checking for?" Andrew asks.

"Let's pull up the client's wish list," Eric replies. They scroll through the document —yet again—to check that their software design reflects the client's wants and needs. Then it's back to the design—yet again—to see if there's anything that needs to be changed or added. A little extra time spent at this initial stage can save much more time, and many more headaches, later on.

> ### SELF-RESPONSIBILITY
> Being self-responsible is about doing the right thing. It's about giving that extra effort, without being told or asked. It's about being mindful of the big picture while digging into the details.

2:30 PM
KEVIN F. & RICK COPPERSMITH
Developers

The programming pair working on "Shropshire," the pair that announced their "final stretch" status at that morning's Stand Up, appear at Kevin and Rick's table. One of them says, "We're looking for an orange dot."

"Do you have an orange dot estimate?" Rick replies.

15 minutes. A perfect billable time interval. Rick and Kevin head over to the Shropshire worktable.

Shropshire involves reorganizing the documentation for a complex program, 80% of which Menlo wrote and 20% of which was written by another company. The documentation for the program—which actually consisted of six separate documents—reflected not only the fact that two companies had worked on the program, but that the work had been done over many years. The client had finally decided that it was time to clean up the mess.

For this project, the client sent the six pieces of documentation over for Menlo to integrate, reorganize and reformat. Someone on the client's end had already tried to do the job, but the effort had caused problems with the content. It had taken several pairs of Menlo developers to address the content issues and get the documentation to what Rick and Kevin are now looking at.

It works. Kevin and Rick okay the orange dot. Now it's ready for a QA review. Definitely a good way to close out the week.

2:45 PM
BRITTANY MORTON & RACHEL CLEVELAND
High-Tech Anthropologists

By 2:45, the Cherubim Show & Tell is coming to a close. Everybody stands up to look over the storycards for the coming iteration. The client team members ask a few questions and throw out a few ideas for possible task changes. The Menlo developers explain how the time estimates were developed and what each proposed change would mean in terms of hours. The discussion raises no significant concerns or disagreements, and by 3:00 next week's plan is locked down.

As the client team gets ready to leave, there's a flurry of friendly goodbyes. The Show & Tell has left both teams feeling satisfied, just the way it was supposed to.

2:45 PM
ANDREW MUYANJA & ERIC COOK
High-Tech Anthropologists

A developer sitting at a nearby table rolls his chair over to Andrew and Eric to ask a quick question about which aspects of a project should be prioritized. The level of interaction between HTAs and developers is high. As the "voice of the user," HTAs function as translators and bridge-builders between users, clients, and developers. Experience has shown over and over again that the high priority Menlo places on user needs, and the specific HTA processes to research and understand those needs, prove invaluable to the end product.

The philosophy at Menlo is that user-focused software is ultimately the most effective for clients. That emphasis raises its own set of questions. Both Andrew and Eric are aware that there are numerous ways to accomplish a given software goal, each with a different price tag. For example, a user might want a search bar, but do they want one that auto-fills possible results? Do they want something that has a drop-down? Each of these elements will take time, and money, to develop, so finding the "right" solution necessarily requires close communication between HTAs and developers.

Moreover, while HTAs try to stay as true to the user data as possible, users are not always consistent. These inconsistencies often make it challenging to draw conclusions about their software needs/priorities. Such cases require the HTAs to interact further with the clients and end users for clarification.

The object, always, is to get it right, to deliver software that best meets the client's needs. So, the priorities matter: "Do we want to do this?" "Let's talk about this." The details matter: "Is this word spelled right?" "Are the text boxes too big?" And always, communication matters.

3:00 PM
VARIOUS MENLONIANS
Walkie

Every day around 3:00, small groups of Menlonians head out the door for a walk around the block—affectionately known as a "Walkie." Today is no exception, and folks from around the room get up, put on their coats, and walk past the bathrooms, up a flight of stairs, and out into an alleyway. It's fall in Ann Arbor and the air is crisp.

As they walk, some talk about work, some about the weekend. Rick says, "A good developer looks both ways before crossing a one-way street." One developer is playing Pokemon Go. Two years after it was big. Programmers.

Many Menlonians are gamers. In less than a month, Menlo is throwing a party in the office where a local area network (LAN) will be established to enable people to play multiplayer video games together. Everyone who wants to participate will bring in their personal computers or gaming systems. Rick is a gamer, but he probably won't attend. "I'll go home, instead. I do want to play Rocket League with everyone, though!" Then he adds, "From home."

Kevin is not a gamer, although he likes to watch. During the last LAN party, he left to browse around in the record store right next door. He's a musician. Eventually, he came back and hung around for a while. "It's just fun to be with everyone," he says. Remember? Kevin is one of the three extroverts at Menlo.

The Walkies group is back at the Factory's back door. On the Menlo-side of the door, there are two sticky notes. One reads, "Keep door shut." The other reads, "It helps keep smells out and dogs in."

They were back to work in ten minutes flat.

3:00 PM
MOLLIE CALLAHAN
High-Tech Anthropologist

When someone called for the afternoon Walkie, Mollie didn't go. She had taken her personal Walkie earlier in the afternoon and now she needs to get the follow-up emails out.

She calls out, "Hey Lisa! Who should I assign the follow-up task to?"

Lisa: "Make it my task, 'cause I'll see it and hunt people down to make sure they do it."

"Thanks!"

Her phone announces a text from her older child, letting her know that he's home from school. She sends him a quick response. Then it's back to HubSpot. She finishes the follow-up on the first lead and moves on to the second.

3:00 PM
RICH SHERIDAN
CEO and Chief Storyteller

Rich is due to meet at 3:00 with Paul, an IT exec who is new to Ann Arbor and wants to start networking to get a better feel for the local IT scene. He's been referred to Rich through a mutual friend. Paul arrived a bit early and he's been examining a copy of *Joy, Inc.* on display just inside the front door.

Rich, just finished with his phone call with Robert, walks over to Paul to greet him. Paul, holding the copy of *Joy, Inc.* in his hand, somewhat surprisingly says, "I'm fortunate to have never experienced fear management myself. But I've known many others who can't say the same thing."

Rich and Paul move over to an open table outside of one of the conference rooms. Paul, who looks to be in his late 40s or early 50s, gives Rich a quick overview of his career, working in reverse chronology. When he gets to his first job, at Hewlett-Packard, he and Rich start trading stories. "My first computer experience was in 1971 with an HP," Rich says. "So, I have a special place in my heart for HP. It was sad for me to see what happened to them."

"I was there!" Paul says. "We were creating great software. But HP was a hardware company. We didn't know how to monetize the software we created."

Rich and Paul exchange stories about people they knew at HP, but there are no overlaps. After HP, Paul had a very successful run at a large financial institution. Then, because of his relationship with his significant other, he ended up in Michigan at his current organization. The conversation drifts back to fear in the workplace.

"Right after I came on board," Paul says, "I started talking with the team. I said I wanted to operate in an open and trusting environment with them. There should be no fear of speaking up. One of the guys talked about where he had worked before, in a company where fear was dominant. He talked about one of their new software releases. Five hundred known defects! Fifty hot fixes in the first week of release! Fear-based organizations create that!"

"Yes, they do," Rich concurs.

As they talk, Paul refers to his experiences at HP, in the financial institution he worked for, and in his current circumstance. In his present role, he's achieved a 75% favorable score on an internal employee survey—an outstanding score in his company. His team is overwhelmingly Millennials. "I'm the oldest, by a long shot," he says. "We just had a successful release. Very few issues." The downside? His company is downsizing, and he's had to let some good programmers go.

Finally, Rich says, "Okay. Let's focus on the positive energy. We need your kind of talent in this community. If there's any way I can help, I'm happy to. What kinds of folks do you want to network with? I have quite a network through LinkedIn, but that includes a lot of people I don't know really well. So, I don't know how helpful that will be. But if you want to connect with someone in my network who's in Ann Arbor, then it's probably someone I know pretty well. There are a lot of start-ups here if you're interested in that sort of thing. Lots of big shops too."

Rich and Paul shake hands and Paul leaves the office. The appointment had run a little long, but Rich is okay with that. He's impressed with Paul's experience, his ideas about management, and his personal demeanor, and he'd feel comfortable recommending him to his closest contacts. It's all part of his commitment to serving the community.

3:10 PM
KEVIN F. & RICK COPPERSMITH
Developers

Kevin, Rick, two other developers, and two QAs, Matt and Art, roll their chairs together in front of one computer. Impromptu clusters like this one pop up all the time, when a larger group of people working on a project need to swap ideas or solve a new problem.

This particular cluster was initiated by the QAs who want to discuss a permissions issue with the software for the Quail project. Art munches on a slice of pizza while Matt raises questions from a notebook.

Rick says, "We're in a good spot again," then takes the group through the work that's been done to deal with the issue. The QAs are reassured, so there's no need to go into fix-it mode. That's good news since the day, and the week, are almost over. Menlo, unlike many, if not most, software shops, treats nights and weekends as sacred, except in rare, dire circumstances.

Michelle, the Quail PM, walks over to tell Rick and Kevin that the other pair of Quail programmers will conduct Tuesday's Show & Tell with the client. She knows Rick has something going on Tuesday morning, so she asks for an update before finalizing and sending out schedules. Rick explains that he'll be in at 10:30, after taking his son in for an eye doctor's appointment. No problem.

3:15 PM
BRITTANY MORTON & RACHEL CLEVELAND
High-Tech Anthropologists

Brian, one of the developers working on Cherubim, comes over to ask Rachel and Brittany a question regarding how to best implement one of the changes discussed at the Show & Tell. Brittany is in the kitchen making a coffee run, so Rachel answers Brian's question. As soon as Brittany is back at the table, Rachel fills her in.

It's crucial for HTAs and developers to always make sure they're on the same page, because what the developers build is based on the designs created by the HTAs, and the insight into end user needs the HTAs gained during their initial work at the client site.

As the afternoon goes on, Brittany and Rachel focus on cleaning up their design in the light of the Show & Tell feedback. Menlo people may not work nights and weekends, but during the day, they work until the "final buzzer."

3:15 PM
JARON VOGEL & GRANT CARLISLE
Developers

Even after their phone call with Night Sky's PM, Jaron and Grant still haven't reached final resolution on how best to deal with the header issue. Grant is still pretty adamant that headers at the top, as opposed to something similar on the side, would make the app look too cluttered. Jaron is growing frustrated, his usually calm demeanor taking on a bit more of an edge.

Grant pushes once again for them to investigate putting what is currently in the headers somewhere else on the page. Jaron comes back with, "Look, I'm not telling you my personal plan, I'm telling you the plan based on the cards. We can't start going off the cards because that adds unnecessary work to the project."

There's a long pause, before Grant says, in a somewhat defeated tone, "Okay. Hopefully, they love it, but I wouldn't be surprised if they don't."

"I agree with you," Jaron says, the edge in his voice now gone. "And that's where my frustration lies." As he has before, Jaron then comes back to the fact that this client nixed the HTA process and Menlo had, albeit reluctantly, accepted the project anyway. "Our HTAs would have had a lot of this laid out in painstaking detail," he says.

"For sure," Grant concurs.

The momentary tension between the two partners is gone, and they go back to work. They'll do it the way it's laid out in the client's mockups, even if they don't think that's the best approach. After all, there will soon be a Show & Tell, and the client can ultimately decide.

3:20 PM
MATT SCHOLAND & ART KLEIN
Quality Advocates

Art and Matt are back to their list of tests for Card 22, the automatic payment/invoice matching function. The permission spreadsheet doesn't express limits, but one category *was* limited on their tests. They've talked over the issue regarding access to information with the developers. They have a sense about what it would take to change permissions. But what would the time estimate be? And is it a Card 22 issue or a bug card issue?

Finally, Matt decides to send an email to the client to see who they believe should have permission. But at this point, without input from the client, he and Art have concluded that this issue would not be part of Card 22. That means Card 22 is finished!

Up at the Work Authorization Board, Art puts a green dot with his and Matt's initials on Card 22. Done! At least until the client weighs in.

Looking for what to do next, they pull Card 366. Matt makes a copy of the card and puts the original back on the board.

On the way back to their table, Matt and Art are stopped by a Dev pair who have some questions regarding their project. Five minutes. (Not that Matt and Art would log it if it were longer. At Menlo, QAs do not estimate their time for billing purposes as the Devs do. Their time is automatically estimated at 10%-12% of each project.)

Card 366 addresses a script that was added to include numerous users. Matt and Art have to make sure it works. They start by discussing the best way to test it, given the data-intensive nature of the software, which increases the variables.

Developing a test that covers all the variables is like solving a complicated puzzle. And solving it? That's the fun part. And they're back at it.

3:30 PM
MOLLIE CALLAHAN
High-Tech Anthropologist

Mollie has been working on the follow-up to the trade show leads, but she has a conference call with a client scheduled for 4:00. James will be on the call too, so she calls over to him. "Are we all set up for the four o'clock or does someone still need to arrange something for it?"

"You don't want me to think of you as a linear person, do you?" James jokingly replies.

"No, because then you'll put me on the PM team," Mollie teases back. With a smile, James says, "No, we're good to go."

Mollie goes back to working on the leads. Once these initial emails have gone out, the next step in the follow-up process will depend on whether, and how, the potential client responds. Will they send another email? Make a phone call? And how long should they wait for a response before a second follow-up? Will she or David or someone else from the Liberty team take over, as she and Lisa discussed? It's an experiment.

As Mollie wraps up the work on another hot lead, Brittany and Rachel walk over to ask her about a question that came up during their Show & Tell. Mollie follows them back to their table to help.

"There was an issue with the food page," Rachel says, "and there was some back and forth about how to solve it, but we didn't know the answer, so we told them we'd follow-up in an email this afternoon. It has to do with how the data is saved."

Quickly, the three of them run through some possible solutions.

"We could use any of those approaches, but we haven't tested them," Mollie finally says. "And don't forget, the test around drag-and-drop functionality didn't work." They agree to send the client several of the possible solutions for further consideration.

Mollie: "Aside from this issue, how do you feel the Show & Tell went?"

Brittany: "I feel that it went well."

Mollie: "Good, I'm glad." She sees James walking across the floor to the conference room for their 4:00 client call. "Have to go," she says.

Mollie hurries to print out a few documents she'll need for the meeting, then heads to the conference room.

4:00 PM
ANDREW MUYANJA & ERIC COOK
High-Tech Anthropologists

Andrew and Eric put their work on hold to join James and Mollie on the Horizons client call. A poster on the wall has a quote from Darwin's *Origin of Species*: "It's not the strongest of the species that survive, nor the most intelligent, but the ones most responsive to change."

Everyone gathers around the small table. James has hooked a monitor up to his laptop, and Andrew sits down to set up the video conference.

"James, can you type in your password?" Andrew asks.

"Oh, if you want, you can just use the little fingerprint sensor in the right corner," James replies, with a chuckle.

"I forgot your fingerprint, James!"

James laughs again and types in his password.

When the link comes up, the sound is working fine and they can hear the client and vice versa, but neither team can see the other. In some companies, this kind of thing would cause embarrassment at best, stress and finger pointing at worst, but no one on either end of the call seems overly troubled by the technical difficulties. "Human suffering as it relates to technology." Even at Menlo.

✦ ✦ ✦ ✦ ✦

The video issue is resolved, and the discussion gets going. These two teams have worked together for some time, and it's obvious that they're comfortable with one another. The conversation is easy, covering how the project is going. The client team describes a few problems they're wrestling with, and the Menlo folks make suggestions.

James asks if the primary stakeholders have offered any insights and feedback. The answer comes back: "No."

James: "Why do you think that is? Are there any positive deviants you can call on to help us understand what needs to be done better?" "Positive deviants" are people who depart from company norms in favor of a better alternative. At Menlo, positive deviants—especially end user positive deviants—are seen as valuable resources, people who should be acknowledged and listened to.

During the meeting, the client shares information and materials that they are by no means obligated to share. In fact, at many organizations, such materials would never be shown outside of the business. In response, James makes it clear that he and the whole Menlo team very much appreciate this exceptional trust, transparency, and commitment to the success of the project.

4:00 PM
KEVIN F. & RICK COPPERSMITH
Developers

> *At regular intervals, the team reflects on how to become more effective, then tunes and adjusts its behavior accordingly.*
> -- A Principle behind the Agile Manifesto

Rick is introduced to his pair partner for the following week. Melissa is in the three-week interview phase of the hiring process. She asks when Rick usually arrives at the office, and when he says between 8:30 and 8:45, she says, with enthusiasm, "I'll be here!"

Wondering about his own set-up for the following week, Kevin checks the schedule, then pumps his arm and proclaims, "K squared next week!" He'll be working with another Kevin.

Menlonians can suggest what they want to work on next and who they might pair with, but Kevin is excited to work with the other Kevin, and Rick is fine working with Melissa. He enjoys the teaching opportunities that Menlo provides him, and says, "I get to teach here more than anywhere else." This, coming from a Ph.D. with a background in academia.

Like many of his colleagues, Rick appreciates the fact that at Menlo there is never only one project to work on or one problem to solve, and he believes firmly that this variety leads to a higher level of productivity and quality. The constant movement from one pair and one project to the next makes for an effective mix of new information and old but still valuable, accumulated knowledge. At Menlo, learning happens in real time, all the time.

✦ ✦ ✦ ✦ ✦

For the first time all day, baby Josiah begins crying, the sound adding a new note to the consistent buzz of Menlo at work. He settles back pretty quickly, and no one is bothered or distracted by his cries. His voice is simply one of many.

Josiah is Menlo baby #22. Menlo baby #1 was born in 2007. So, Menlonians are used to having babies around. Even if there are no infants in their private lives, they are sure to run into them at work.

Rick's son, Billy, was never a true Menlo baby. He was a one-day interviewee who didn't pass his father's test. "I brought him in once, and it took me half an hour to change his diaper and feed him, and it was not easy to get any work done at that point." Had Rick decided to continue bringing Billy to work, he would not have been the first, or last, dad to do so.

Billy is now two years old, and he's been on a waitlist for daycare in Ann Arbor for two and a half years. This is one of the reasons for Rick's biggest complaint about Menlo, that the company is located right in the center of a town where housing is extremely expensive and childcare very limited. For many Menlonians, the only option is to live well outside of the city and endure a tough commute to get to work every day. Rick says, "I have this vision of opening a daycare center so that it's easier for parents to work at Menlo."

As it continues to grow, Menlo has to attract and retain top talent, and at least in part that will depend on finding practical solutions to problems such as access to housing and childcare. As he drove around earlier in the day looking at possible sites for a new, larger Menlo Factory, Rich Sheridan had been looking for those solutions.

4:15 PM
RICH SHERIDAN
CEO and Chief Storyteller

Rich is back at his table. Michael, the young programmer who talked to him at lunch about hiring Vaughn, stops by. Vaughn is with him.

Rich reaches out his hand and gives Vaughn a hearty "Congratulations!"

"Thanks," Vaughn replies. "I'm very excited."

✦ ✦ ✦ ✦ ✦

Rich has an email from Jane, with the weekly finance report and an attachment on the status of Accounts Receivable. The email concludes, "No concerns," which means Rich doesn't have to drill down into the data. If Jane is okay with everything, so is he.

✦ ✦ ✦ ✦ ✦

Rich looks around and doesn't see Anna, who had been working at a nearby table, so he falls back on a quick email to tell her about his meeting with Paul.

"Hey Anna, I just met someone who seems like a good guy, an IT exec with a solid background who really understands the downside of fear-based management. I'd like to help him network with like-minded folks. Plus, at his company, they're downsizing and he's had to fire a lot of good programmers. I'd like to invite him to our December Deep Dive program, at our expense. Okay?"

Deep Dive is a weeklong immersion program for people who want to learn about Menlo's culture. Having Paul attend might be a good investment, given how he thinks about leading. And the fact that he knows a bunch of good programmers who could be looking for work. Finding IT talent is always a top priority for Menlo, but even more so now when client demand is putting a strain on current resources and the competition for talent is so intense. You never know what might come of bringing someone like Paul into the Menlo network.

✦ ✦ ✦ ✦ ✦

Rich walks over to where Lisa is standing. "Josiah was certainly mellow today," he says, peering at the little guy who is peeking out of his carrier.

"He did a bit of chirping," Lisa replies, with a smile.

Rich: "He's just vocalizing his excitement about life!"

4:20 PM
BRITTANY MORTON & RACHEL CLEVELAND
High-Tech Anthropologists

Brittany and Rachel have finished cleaning up the Cherubim designs, based on the client feedback at the Show & Tell. With the day, and the week, nearly over, they slip into casual conversation. Rachel—glancing over at the nearby Levels Board—mentions that in the short time she's been at Menlo, she hasn't had time to get familiar with it. Brittany offers to take her through it, so they take a walk over.

As she explains the various compensation levels, and how the sticky notes indicate where everyone in the company stands, Brittany says, "I have no problem with my salary being displayed for others to see! I wish this were a more widespread practice." Her comment prompts a discussion about how Menlo's transparency about compensation helps prevent discrimination based on things like gender, race, or sexual orientation—discrimination that's all too common in other workplaces.

As far as actual compensation, Rachel and Brittany both agree that you could probably find some local companies that pay better. But Menlo's recent pay increase had helped close that gap, and besides, Menlo is just a great place to work. The high level of transparency creates an atmosphere of real trust. There are constant opportunities for growth. And then there's work-life balance, the freedom to leave your work at the office, with no weekend work or taking your computer home to work at night, even allowing people to bring their babies to work—that all has real value!

✦ ✦ ✦ ✦ ✦

> **CARING**
>
> In high-performing organizational cultures, people are sensitive to the feelings of their colleagues. Caring is an expression of our full humanity; it belongs at work, and it actually makes for a more productive workplace.

As they wrap up for the day, Rachel asks Brittany for feedback on how she did in the Show & Tell. Brittany tells her that she did fine, and as the conversation broadens, Rachel shares the broader insecurity she's feeling about perhaps not succeeding in her new role. Brittany is supportive and reassuring, making a sincere effort to reinforce the human connection she and Rachel have been building as they've worked together.

Menlo is not a perfect place, and the people who work there are not perfect either, but the Menlo culture does encourage and enable these kinds of very human interactions, this kind of empathy for one another and willingness to support one another—and that makes it a special place.

4:44 PM
KEVIN F. & RICK COPPERSMITH
Developers

Kevin and Rick have published the code they've been working on to the database.

A dog, not a Menlo canine, comes running into the office from somewhere else in the building, its owner in hot pursuit. It ends up at Kevin and Rick's table. They get hold of its collar and give it a back rub until the owner catches up. Apologies and laughter. A nice way to end the day.

Matt and Art, the QAs who've been reviewing the storycard Kevin and Rick had been working on earlier come by to point out a couple of errors. Kevin and Rick take a crack at fixing the errors, and the tests they run are *mostly*, but not completely, successful. With their eyes on the clock, and with Rick having to leave at 5:00, they shelve the card until Monday.

Ideally, either Kevin or Rick, working with their new partners, will finish the card next week. If for some reason, that can't happen, they'll leave some "Dev notes" on the card so some other pair can pick up where they left off. If that happens, there will undoubtedly be a quick "Hey Kevin," or "Hey Rick" to ensure a smooth, successful hand-off.

4:50 PM
MOLLIE CALLAHAN
High-Tech Anthropologist

As five o'clock grows closer, Menlonians return their used coffee mugs to the dishwasher, and generally tidy the place up in anticipation of the weekend.

✦ ✦ ✦ ✦ ✦

In the conference room, Mollie, James, Andrew, and Eric are still on their client call. All have taken copious notes during the meeting, and it's clear that they have a lot to think about in the coming days and weeks. Five o'clock comes and goes, but the meeting is productive, so no one seems bothered. Finally, at 5:20, good-byes are exchanged, and the video link is shut down. The Menlo folks straighten up the room and head back to their desks.

By now there are very few people left in the office, and the buzz that filled the room all day has been replaced by a gentle hush.

5:00 PM
MATT SCHOLAND & ART KLEIN
Quality Advocates

The week is over. They did some good work, and there will be more puzzles to solve next week. But now, it's time to enjoy the weekend!

5:00 PM
LISA H.
Project Manager

By 5:00, the Factory is emptying out. Lisa gets pulled aside by someone from another project to discuss a developer who has been at the same level for quite some time. They agree to talk with the team, observe the developer, and get a better handle on what's going on.

Lisa moves to the Work Authorization Board to update a couple of storycards. While she makes the changes, she chats with another PM about a Project Management class they will be leading together.

Having posted next week's work assignments for her team, Lisa retrieves Josiah from the colleague who's been holding him. Throughout the day, Josiah has stayed mostly under his Mom's care, in his kangaroo pouch carrier or on a blanket on the floor beside Lisa's table (although she did put him down for a long nap in the crib in the baby room). While a few people have held him for a few minutes, Lisa feels that it's the parent's responsibility to take care of their babies during work hours. So, only when people are winding down at the end of the day does she feel comfortable allowing them to get their baby fix. There are always plenty of arms waiting to hold him, and Josiah seems perfectly content to accept last minute hugs from his Menlo family.

5:00 PM
MARILYN MacEWAN & McTAVISH
Office Assistant and Office Canine

While Lisa finishes up, Marilyn and McTavish prepare for their hour-and-a-half commute home. McTavish circulates through the Menlonians standing around with backpacks on ready to head out, through those who are choosing books to take home from the library shelves, and those who are closing out their time sheets and shutting down their computers. At every stop, he gets a final scratch or pat, his tail wagging in appreciation. He's had another good week.

Marilyn finally rounds him up, puts his leash on, and together they head to the car for the long ride home.

5:04 PM
KEVIN F. & RICK COPPERSMITH
Developers

The best laid plans…

Kevin says to Rick, "It's 5:04. You have to go."

Rick quickly gathers his things and heads for the back door.

Kevin stops to chat with a few coworkers before he too, makes his way home.

5:05 PM
ANDY WALTERS
Developer

As everyone is heading home, Andy, a developer who has been working at a client site, is just coming in. He could have just gone home, but he wants to check in with James to tell him about what he learned that day.

At Menlo, the employees are empowered to run their own schedules, to determine how and when they want to work. Everyone trusts everyone else to get the work done.

As James likes to say about the Menlo philosophy: "Work hard when you're here, then go home."

5:20 PM
ANDREW MUYANJA & ERIC COOK
High-Tech Anthropologists

The Horizon client call went well. It was part brainstorming session, part status update, and both teams left with some defined, actionable next steps. Everyone was an active participant and contributed to the plan for next steps.

Eric has plans at 5:30, so he heads out right after the meeting. Andrew walks back to their table to review some of the changes they had made earlier. He feels okay doing it without Eric, since they are both aware of what's going on and what still needs to be done. That said, he doesn't stay long; at 5:45, he puts on his jacket and walks out of the office.

5:28 PM
RICH SHERIDAN
CEO and Chief Storyteller

Asha has finished sprucing up Rich's PowerPoint deck. She walks up to his table, says, "Can you check through this?" and loads the presentation onto his computer.

The slides look great, except for one image—a picture of a broken heart meant to illustrate the stage of his career when Rich had lost the joy he once had in his work with software. By changing the color palette, Asha has ended up with a black heart, which makes Rich wince. It takes Asha less than five minutes to make the heart red.

Done.

5:30 PM
MOLLIE CALLAHAN
High-Tech Anthropologist

> Fear does not make bad news go away. Fear makes bad news go into hiding.
>
> —Richard Sheridan

A poster on the wall at Menlo Innovations

"I have an idea for the upcoming speech in November," James says to Mollie. "It came to me during the Horizons call when we were talking about their problems getting people to help each other out. The topic for the speech has to do with fear at work. So, what if we include the fear that 'if I help everyone else with their work, how will I get *my* work done?' with the obvious answer being that everyone will help you in turn. What do you think?"

"I like that," says Mollie. "I know the PMs are worried about growth opportunities for the team and managing that with our existing workload. This morning's conversation after Stand Up is a great example of that."

✦ ✦ ✦ ✦ ✦

As James walks off, Rachel comes up to talk to Mollie. "What time should we come in on Monday to prepare for our Kick-Off meeting?" she asks.

The Kick-Off meeting is a time for a team, which might include some new members depending on how the pairs were shuffled, to share what they learned from the last Show & Tell and plan out the week's work.

"How familiar are you with the project?" Mollie says, knowing that Rachel didn't have much experience on this project and might need more time to get caught up.

"Fairly."

"Then I think 8:30 would be early enough. That way we have enough time to bring you up to speed and get on the same page."

"Sounds good to me. See you then. Have a great weekend!"

✦ ✦ ✦ ✦ ✦

Back at her worktable, Mollie checks to make sure she's really ready to head out for the weekend. At the adjacent group of tables, Rich is talking to James about Project Liberty, and he calls Mollie over to

A Kick-Off meeting to plan the week's work

ask how the Liberty team will decide whether he or James needs to follow up with someone.

"We're using HubSpot to log leads, contact information, and a log of interactions with a lead or company. The program also lets us assign follow-ups to someone on the team. In some cases, you or James will clearly be the best person to reach out to the lead, but in other cases it might be one of the HTAs or back-office team. Lisa has overall responsibility, and she'll consult with the team to decide who would be the best person to actually make the contact—which again, could be you or James. Then Lisa will stay on top of the process to make sure the follow-up actually happens, and next steps are planned."

✦ ✦ ✦ ✦ ✦

Satisfied with the explanation, Rich and James continue their conversation and Mollie returns to her table to quickly upload the context for the last of the three hot leads. Satisfied with what she accomplished today, she packs her oversized bag, returns her coffee mug to the dishwasher, grabs her coat, and walks out the double glass doors at the front of the office.

5:30 PM
JARON VOGEL & GRANT CARLISLE
Developers

By 5:30, almost all of the Menlonians are on their way home. Grant and Jaron have been thoroughly absorbed in the task at hand, testing some of the elements they introduced to format the headers, so they've barely noticed the time. They've basically been at this for eight hours.

Jaron looks up, and says to Grant, "I think we should wrap up for the day. I want to make sure we have enough time to do your second week review!"

"Oh yeah!" Grant replies. "I think you're probably right." As nonchalant as he sounds, Grant is actually incredibly nervous. Jaron calls Al over, as well as Andy, not only because Andy worked with Grant last week, but also because he is one of the most senior employees at Menlo.

They quickly form a circle, and just like that, the review is underway. Andy starts things off with something to put Grant at ease. "So, Grant, how do you feel at the end of your second week?"

Grant: "I think this week went really well! I've said from the beginning that I love it."

His response is met with smiles. Then Andy, in a bit more serious tone, says, "One thing I noticed while working with you during week one, is that you seem pretty new to unit testing. That's why I made you drive a lot more during the writing of those unit tests."

"Yeah, Jaron has been making me do that too," Grant responds.

With a bit of mischief in his eye, Andy says, "Yeah, well secretly, we've been colluding behind your back this whole time." Grant has suspected as much, but he can't help but shoot a smirk in Jaron's direction.

Al jumps in with, "My struggle right now is that I don't know what level pay grade to put you at."

"Yeah," Jaron adds. "If we bring you in at some level above Associate, the project manager expects that they can put you with someone new. Our struggle is, if we pair you with someone new, would you two be able to succeed in a unit test-driven environment?"

"Would I be able to lead us through that?" Grant responds, nodding his head.

"Exactly," Jaron says.

Al jumps back in. "That being said, you are also incredibly valuable when it comes to discussing architecture. You are clearly knowledgeable and provide great discussion at Kick-Off."

"When usually it takes a year for our associates to even say anything at Kick-Off," Jaron adds.

"Exactly. So that's why we're struggling with where to bring you in," Andy says.

"Well thank you," Grant began. "As far as unit testing, I'm definitely on board that it is what we do. That it's how we work at Menlo. And I'm willing to put in the time to learn how to do it better."

"That's great! That's what we like to hear," Jaron says. "I'd like to see you take the keyboard and try to write a test that you're uncomfortable with. I shouldn't have to ask you to because not everyone you pair with might."

Grant nods his head in agreement. He knows that this is his weakest area and while he made improvements during his second week, he also knows that he needs to learn more.

Standing up, Jaron motions for Grant to follow him to the bookshelves that line a portion of the walkway leading in from the main doors. Andy and Al walk over too, and Al plucks one from the bottom of the third shelf, The *Art of Unit Testing*. All three of the Menlonians share how this book was an awesome resource when they first started learning about unit testing. They hand it to Grant and suggest that he leaf through it over the weekend and take a crack at some of the practice problems.

"You clearly have the ability to learn how to do this well," Al says. "You just need to put some time into becoming more comfortable with it."

Grant takes the book with a smile and puts in his backpack. He's pleased with the way the review went and determined to take his game to a whole other level next week, the last week of his pre-hiring process. With good-byes all around, he slings his backpack over his shoulders and walks out the door. His weekend will definitely involve spending some quality time with *The Art of Unit Testing*.

5:48 PM
RICH SHERIDAN
CEO and Chief Storyteller

Rich stops Andrew who was about to leave. "I had a good talk with a guy today. He just laid off 24 developers."

Andrew laughs. "Any we can use?"

5:58 PM
RICH SHERIDAN
CEO & Chief Storyteller

Rich gathers up a few stray coffee cups. His day is going to end where it started. At the dishwasher.

The day had not been unusual. As Menlo's public face, he had networked with colleagues outside of Menlo, and learned some interesting things along the way, filing them away for the future. He had checked emails. He had raised some issues about a client, looking for a win-win. He had paired with an HTA to develop a slide presentation, one that would hopefully lead to more folks knowing about Menlo and eventually lead to more work. He had mentored. He had connected with someone who needed help, and who might ultimately be of help. He had spent time with a Menlo baby. He had put dirty cups in the dishwasher.

An ordinary day, but one that demonstrated the "mundane magic" that is Menlo Innovations. A day of joy.

> People need joy quite as much as clothing. Some of them need it more.
>
> —Margaret Collier Graham

A poster on the wall at Menlo Innovations

SATURDAY
THE MENLO INNOVATIONS FACTORY
_____

The lights are off. No one is in the office. It is that way all day.

AFTER FRIDAY

Rich Sheridan's new book, *Chief Joy Officer*, was released on December 4, 2018. As expected, the book gradually built momentum for Rich as a speaker. He quickly had inquiries from Denmark, Germany, Ireland and Qatar. He accepted the speaking invitation from Barry-Wehmiller's CEO Bob Chapman to be one of the keynote speakers at the 2019 conference of the Advanced Manufacturing Engineering Conference. Later in the year, he also spoke at the Center for Automotive Research, the Association for Advancing Automation, and at the TriAgile Conference.

Menlo also expanded its speaker's bench and started involving more Menlonians in giving public talks, particularly locally. A career development opportunity much appreciated by many.

✦ ✦ ✦ ✦ ✦

The Liberty Project continued to generate new business. The staff grew to over 50 in about six months. A few Menlonians left. Rick Coppersmith went to work at an IT security company. Anna Boonstra jumped on a good opportunity at Ford. Marilyn MacEwan, long-time Menlonian, retired.

✦ ✦ ✦ ✦ ✦

With Marilyn no longer at Menlo, McTavish was no longer the unofficial Menlo canine. That role was taken over, in true Menlo style, by a pair of dogs, Rambo and Dexter, who were under the care of Giovanni Sturla and Rich Sheridan, but who happened to be owned by neither.

✦ ✦ ✦ ✦ ✦

Grant Carlisle successfully completed his third week of the intensive interview process. So did Melissa, who was going to be paired with Rick Coppersmith the week that followed Friday. Both were of the 60% of applicants who complete their three-week engagement and got a full-time job offer from Menlo. Both accepted, as do most such applicants who make it through the intensive hiring process that is Menlo's.

✦ ✦ ✦ ✦ ✦

One of the four projects highlighted in this book ended before completion. The client PM decided he wanted a second outside team to take over. That rarely happened at Menlo, and sometimes when it did, the client ended up coming back.

The other projects highlighted in the book all came to successful conclusions, meeting all milestones set by the client. In one case, the client PM decided to move the project along for a while internally to get his team deeply knowledgeable about how the software functioned. He thought it likely he would return at some point for another shot of Menlo creativity. In another case, the client was so satisfied with the work that they moved to their "start-up launch phase" and rented space in the Menlo Garage.

✦ ✦ ✦ ✦ ✦

The problematic project that Rich, James, and Pat spoke about over lunch remained problematic. Differences in style between Menlo and some members of the client team became more apparent, but the Menlo team continued to find workarounds.

✦ ✦ ✦ ✦ ✦

Talk about suffering as it relates to technology. As the saying goes, "To err is human, but to really foul things up, you need a computer." On February 3, 2020, the Democratic Party in Iowa had the great misfortune of trying to aggregate local caucus results using a new app designed for mobile phones. Some bug, and some lack of testing at scale, created massive egg-on-the-face for Iowa Democrats. In the old paper-and-pencil days, the Iowa vote might not be known until early the next morning after the caucuses closed. But thanks to the new app, it was days and days before the votes were all counted.

Although that foul-up was embarrassing, it wasn't tragic. But on March 10, 2019, Ethiopian Airlines flight 302 crashed after take-off from Addis Ababa killing all 157 people on board. This crash was the second such incident after the crash of the Lion Air flight 610 that killed all 189 passengers and crew on October 29, 2018.

Both crashes involved Boeing 737 Max planes equipped with computerized flight software called the Maneuvering Characteristics Augmentation System (MCAS) that automatically adjusted the mechanical systems on the plane when it sensed the plane was going too fast or too slow, given the angle of the nose of the plane. The inadequate training of pilots using this new application, but more fundamentally the programming of the CMAS itself, was the source of the problem. This was truly a "software emergency," one that resulted in hundreds of deaths, the grounding of all 737 Max planes, and a disastrous blow to Boeing's stock and reputation.

Rich Sheridan commented that, in situations like this, the tendency for the manufacturer was often to blame "pilot error." This is akin to the "stupid user" excuse software developers often fall back on when their products don't work as well in the real world as they did back at the development shop.

Menlo Innovations isn't into stupid user excuses. That's why they invest so much in the High-Tech Anthropologists who work hard at understanding the user experience in the real world. It's also why they invest so much in their Quality Advocates. More than anything else, Menlo never wants to be responsible for a catastrophic software emergency.

✦ ✦ ✦ ✦ ✦

The Coronavirus pandemic disrupted the normal workflow at Menlo. As the pandemic hit first China and then Italy, the Menlo team began thinking about what they could do if drastic measures were called for in the U.S. For an organization whose entire culture was aimed at fostering close collaboration, via such practices as pair partnering and Daily Stand Ups, there obviously would be serious challenges. It was time to "run the experiment."

Menlo did have some experience with remote work. Kealy had worked remotely on a temporary basis when she stayed with her parents for a time so she could provide health care for her dad. Randy, in Moscow, was a more permanent experiment in remote pairing. And Menlo teams regularly used remote links to stay connected with client teams, including using them for virtual stand ups.

When the time came, the whole team went home and connected via team-supporting software. Developers took Dev boxes with them, with one member of each pair having the computer physically in front of them, and the other virtually plugged in so that either partner could drive or navigate. Project teams were connected in separate chat rooms to enable conversations that went beyond a specific pair on a project. Daily Stand Ups still took place via video, with pairs or individuals entering a chat message to specify they wanted the virtual Viking helmet to be able to go next. HTAs experimented with virtual end user research.

Whether the team eventually returns to being completely co-located, or remote work becomes the new normal, or Menlo evolves to some combination of the two—that remains to be seen.

✦ ✦ ✦ ✦ ✦

On a happier note, Baby Josiah left his mother's baby front carrier and headed off to daycare. Menlo Baby #23, came along in January, Menlo Baby #24 arrived in the spring of 2019, and Menlo Baby #25 joined the family in September.

✦ ✦ ✦ ✦ ✦

Menlo Innovations is still targeting significant growth within the constraints of preserving their unique culture. They have hired a Chief Strategy Officer, an accomplished veteran from the non-profit sector, who has particular strengths in marketing. She quickly introduced Menlo to *The Four Disciplines of Execution*, and the term "4DX" entered the Menlo vocabulary.

Rich and James are in brainstorming mode with some real estate developers on a potential future, larger Menlo Innovations site.

We wish them good fortune!

EPILOGUE

Michael Pacanowsky

I was first introduced to Agile approaches to computing by Steve Denning. Steve had written a book *The Leader's Guide to Radical Management: Reinventing the Workplace for the 21st Century*. Steve, who had a storied career at the World Bank, got interested in Agile as he was looking for ways to dramatically improve the generally poor performance of so many bureaucratic organizations. Agile approaches to programming seemed to offer a rich set of practices that could be applied by any organization to adapt more quickly and effectively to external trends and exigencies.

Steve was on the Board of Directors of The Scrum Alliance, a non-profit training and certification organization in the programming space. He had the idea to go on a Learning Journey, and bring along a cadre of folks who would join him in "search-and-study" excursions to various organizations that were employing to a greater or lesser extent Agile approaches to management. Most of the site visits were to companies in the software industry, but not all, since many Agile practices had migrated from software development to lean manufacturing or to other industries.

Before returning to academia at Westminster College, I had had extensive working experience at W. L. Gore & Associates, an organization well known for its unique culture and its record of delivering extraordinarily innovative products, and Steve invited me to be one of his fellow travelers. It was through my time with Steve's Learning Journey that I learned about Agile and that I met Rich Sheridan.

We visited eight or nine companies where Agile practices were having a significant impact. Most of the time, we focused on how the companies actually handled various practices. "When do you do stand ups? How long do they last? How many people participate?" "How long are your sprints?" "How do you get the 'voice of the customer' into your development planning?" "What's been your experience with pair programming?" But as we asked these questions, with increasing frequency, the leaders of the organizations we were visiting would say something about how changing past practices required first a change in mindset. That is, just holding 15-minute morning stand ups without a fundamental change in the way people thought about their work wasn't a winning strategy. To make Agile work, you had to first move from a bureaucratic mindset to an entrepreneurial mindset. You had to move from "I do what I do because I want to please my boss" to "I do what I do because I want to please our customers." *Before you could effectively change how you worked, you had to change the way you thought.*

For me, coming from Gore, this concept of mindset was just another word for culture. For me, culture *was* mindset. It was how we interpreted and made sense of what was going on around us. It was the interpretive structures and processes that had us see certain things, to feel in certain ways about what we saw, and to think through and then act under those circumstances. But culture was more encompassing than mindset. One of the Learning Journey participants who seemed to be in sync with my thinking about culture was Rich Sheridan of Menlo Innovations. Indeed, when we finally had the chance to visit Menlo, what jumped out was how the Menlo culture was comprehensive, and how their various practices, principles, values, and deep-seated assumptions about work and people were tightly interlinked and mutually reinforcing.

When I left Gore for Westminster College, I had planned on reflecting on my Gore experiences and the lessons learned so that I could bring some of Bill Gore's thinking about people and work and organizations into the Bill and Vieve Gore School of Business. While I had been at Gore, many of the projects I worked on were culture-related, and I believe I was often seen as one of the spokespersons for the Gore culture—both internally and externally. When I got to Westminster, through extensive reading of academic and business thinkers and through contact with leaders like Rich Sheridan, I ended up having a richer set of exemplars to draw upon. My goals expanded. I didn't want to just understand what made Gore and its culture work so well. I wanted to understand, generally, the deep dynamics of any high-performing organizational culture.

As a result of that effort, my colleagues and I at Westminster's Center for Innovative Cultures ended up developing a model of the "Thriving Organizational Culture" and elaborated those ideas in a book called *The Thriving Organization: An Exploration of the Deep Dynamics of High-Performing Organizational Cultures*. In this book you are now reading, which takes you through a day-in-the-life of Menlo Innovations, we've dropped in some signposts that point to places where you can "see" culture happening at Menlo. What I want to do in this epilogue is lay out the model in a bit more detail, show how Menlo Innovations manifests in an interlocking way the various cultural elements of the model, and draw out some of the implications as you think about the culture of Menlo or, perhaps even more important, that of your own organization.

✦ ✦ ✦ ✦ ✦

At the outset of this book, we said we wanted to do three things:

- Inspire you to imagine new practices you might take to your own organization.
- Give you the confidence that these ideas can often be pretty easy to implement.
- Provide you with a deeper understanding of how organizational culture works and why understanding it is so crucial to successfully introducing positive change in your organization.

I hope by now you've come across any number of new ideas from Menlo that you can imagine might be helpful in your organization. I also hope that you see that what Menlo does isn't rocket science or require other-worldly beings to make it happen. What they do at Menlo on a daily basis is, in some ways, pretty mundane. But there is a mundane "magic" at work at Menlo, and it is their culture.

So now I want to tackle the third objective of the book, and give you a deeper understanding of how organizational culture works—and specifically how it works at Menlo Innovations—so that you can have a better chance of successfully making positive change where you work.

✦ ✦ ✦ ✦ ✦

First, a few words about what we mean by culture. If you read any of the popular or academic books and articles focused on organizational culture, you will rarely see the term defined. There seems to be an assumption that we all know what an organizational culture is, and we all mean the same thing when we use that term. But do we?

In fact, when I ask students, workshop participants, or conference attendees how they define "organizational culture," I get quite a varied set of responses. There are themes, however. Some people have a behavior view of culture. It's our behavior. It's what we do. It's how we do things around here. One wag defined culture as "what someone does when the boss isn't around."

Others offer a behind-the-behavior view of culture. It's our values. It's our norms. It's our beliefs. It's our habits of mind or habits of heart. Culture precedes behavior and causes it or influences its expression. Still others offer a more expansive view of culture. It's our identity. It's our personality. It's the "smell of the place." It's our brand. And finally, there are some who—I would argue wrongly—equate culture with the perks available or lack thereof: contributions to a 401K, beer busts, unlimited PTO, paid maternity and paternity leave, free food, and so on.

Except for the perk-based view of culture, there is a sympathetic relationship between the various other definitions. Culture is what we do. But culture is also how we think about things that influence what we do. And it all sort of hangs together as a seamless whole. So, we can generally continue to navigate conversations about organizational culture without agreement on definition because most people's definition is close enough to others' that we don't actually disagree.

But I think there is a more productive, more powerful, definition of culture. That definition comes from Edgar Schein, the academic grandfather of organizational culture. Ed has been writing about organizational culture at least since the 1970s. He derived many of his insights about culture from his intensive and long association with Digital Equipment Corporation—a company that no longer exists but in its heyday was a computer industry pioneer and a "place to be." Ed's classic book *Organizational Culture and Leadership* was first published in 1985. In 2016, Ed and his son Peter published the 5th edition.

According to Schein, organizational culture is "*a pattern of basic shared assumptions that a group or organization has learned as it solved its problems of internal integration and external adaptation that has worked well enough to be considered valid and is passed on to new members as the correct way to perceive, think, and feel in relation to those problems.*" When I ask people what they think of the definition, most ponder it, and some say, "Wow! That's long!" And it is long, certainly longer than "our personality" or "our values." But what's important is the understanding it provides of what culture actually *is* and *how it comes to be*. Let's tease out some of the key implications of Schein's definition.

First, culture is not one "thing," like values, or norms, or even behaviors, but a pattern, a set of interrelated assumptions, interpretations, and understandings that provide us with a way to perceive, feel, think, and act. Schein layers these different cultural elements into two main groups: those that are visible and that we can see or hear when we walk into an organization (he calls these "practices" and "artifacts"), and those that are not visible, but are the structural supports for the practices and artifacts that make what we see and hear meaningful and sensible (these he calls "values" and "assumptions").

Think of it this way. If you go into an organization and you see a manager holding a performance review with an employee, and the session is taking place in the manager's office with the manager behind his desk and the employee in a chair across from him, this arrangement isn't random. It is built on certain values—perhaps "respect the boss"—and assumptions—perhaps, "the boss has power and information the employee doesn't." Given those values and assumptions, this event as you see it "makes sense."

But suppose you go into Menlo and you see their equivalent of a performance review session—a bunch of people seated around a table all chiming in with feedback on the strengths and areas for improvement for the Menlonian being reviewed. This arrangement is not random either. But it's not the same as the traditional review because it's based on different values and different assumptions. Here the values may include "respect the employee" and the assumptions may include "the team has the knowledge, and it should use this knowledge to empower the employee."

Second, culture is not an aspiration. It is not what we would like to be, but what we actually are. A culture poster, a slide deck the CEO uses when she onboards a new hire, a statement of "what we believe" on our website are all, in Schein's terms, cultural artifacts. But they aren't the culture. The culture is revealed by how, and how much, these artifacts are actually manifested in behavior. Do we walk our talk? If so, then our culture is one that takes its aspirations seriously. But if we don't walk our talk, and sadly that is often the case, then our culture is one that encourages hypocrisy. We let everyone know (tacitly, not explicitly) that we don't mean what we say and it's okay.

So, if culture isn't laid down by the CEO in some statement of beliefs, where does it come from? Here I think is one of the most important implications of Schein's definition. Culture comes about over time as an organization grapples with challenges and either succeeds or fails. If they succeed, then however they perceived the challenge, however they felt about it and thought about it, however they responded to it—those perceptions, feelings, thoughts, and actions are likely to be retained. When they face a similar challenge, they will draw upon what worked before. On the other hand, if the organization fails to successfully meet the challenge, then the related perceptions, feelings, thoughts and actions are likely to be abandoned in the search for a new approach. Or we may persevere with what doesn't work, and eventually our organization will collapse.

Schein says that these challenges come in two sorts—challenges of internal integration and external adaptation. What he means is that every organization has to manage two dynamics—how do we work together and how do we manage the constraints the outside world puts on our capacity to deliver value. As an organization evolves, it must constantly be developing new routines for dealing with changing internal and external challenges. An organization that is small starts to grow. How does it manage challenges of internal integration? Who do we hire? How do we onboard them? Do we promote a cooperative spirit or a competitive spirit? What do we do with folks who don't perform? What do we do with folks who perform but rub people the wrong way? What do we do when we get to be 100 folks, or 1000 folks, or 10,000 folks? Do we just keep adding more offices and space to the shop floor? Do we start new sites in other locations? In other countries? All of these are questions of internal integration.

Similarly, over time, every organization finds that its external environment changes, and the organization has to decide if it will maintain the status quo or change and try to adapt. How do we compete with low-cost competitors? How do we handle new government regulations? How do we respond to a new product offered by a competitor? What do we do when the markets we sell into are suffering financially? What do we do when the markets we sell into are booming? Again, we try things. If they work, we are more likely to try them again when a similar situation arises in the future. If they don't work, we are more likely to try something new.

A most important implication of this notion of culture evolving as a consequence of successful adaptation to challenges is an understanding of the importance of the Founder on the culture of an organization. How does an organization act, once it is launched? Usually, the actions taken are an expression of the Founder's insights, preferences, and personality. The Founder does what feels and seems right to him or her. If the Founder trusts his or her gut in making a decision, that will become a template for future decision-making. If, on the other hand, the Founder believes in looking at all sides of an issue before making a decision, even ensuring that contrary points of view are expressed, that will become a template for making decisions. And so it goes. Who does the Founder hire? How? That becomes a template. What strategic vector does the Founder have an affinity for—innovation? efficiency? close-to-the-customer? That will become a template for strategy. (Interestingly, one of Rich Sheridan's key drivers for founding Menlo Innovations wasn't just his passion for well crafted, working software. It was that he wanted to keep working with James Goebel—his main collaborator in all the innovations he was trying at Interface Ssytems that would be the foundations of Menlo. It was the sheer joy of collaboration that Rich experienced, loved, and wanted at the heart of his new organization.)

As Schein says, these cultural templates last because they are deemed valid. They are tried and true. You can hear the Founder now. "We've been successful because we made good quick decisions based on gut feel, not deep analysis. We hire folks like us, comfortable with deciding quickly and not pussyfooting around with all that analysis. And we really really know our customers and what it takes to delight them. That's why we're successful!"

As long as the company *is* successful, then these templates become the established routines for action. If it starts to fail, however, then someone—maybe the Founder or maybe someone the Founder trusts—will say, "This isn't working, we have to try something different." And that's how culture evolves. (Of course, if the company fails, no one really cares about its culture, except as a cautionary tale.)

A couple of other quick points. It should be clear that there is an interdependent relationship between our cultural routines for internal integration and our cultural routines for external adaptation. What happens in the outside world may require us to change how we do things internally. Similarly, how we do things internally may position us in better or worse ways to address external changes. But, unfortunately, many people think of culture as something that has to do only with internal integration—how we work together—rather than the broader context that includes how we adapt to the outside world. That's why, too often in too many organizations, culture is the purview of the HR department, whose initiatives are strongly (or passive aggressively) resisted by line managers who have to deal with constantly changing external demands.

Finally, it should be clear from this discussion of Schein's view of culture that each organization has its own unique culture. That is, each organization has grown up with its own unique set of challenges of internal integration and external adaptation. Some have grown slowly; some have grown rapidly. Some have created value through innovation. Some have created value through efficiency. Some have anchored themselves locally. Some have become global players. In each case, they have developed their own unique patterns of assumptions and interpretations and understandings whereby some actions make sense and other actions make no sense.

That's why anybody who moves from one organization to another, at least anyone with any savvy, almost always begins by trying to ferret out what's different about their current organization from what they experienced in their past organization. What are the trip wires that can derail me? What are the leverage points that can lead to success? Do I speak up in a meeting if I have a different point of view from what the boss just said? Do I speak to my boss privately after the meeting? Do I just keep my mouth shut?

Now, because each organization has a unique culture and a unique way of understanding what it makes sense to do, it is sometimes very difficult to graft a best practice from one organization onto another. If you don't have the supporting infrastructure—the deeper cultural elements that make the practice work—then a good practice from some stellar organization may be a disaster for you. You may have to examine your cultural infrastructure—the values and assumptions and patterns of understanding—to see if yours are aligned in such a way that they will support an imported "best" practice.

James Goebel tells a story about what sometimes happens when visitors tour Menlo. (I suspect he embellishes a bit.) "They come in and are blown away by what they see—the energy, the collaboration, the speed. So, they get all excited about wanting to bring Menlo back to their organization. And they ask us for things like the specs on the tables and chairs we use. And they want to know exactly what corkboard we use for our Work Authorization Board. And they take pictures of the posters. Then they go back to their organization, tear down the cubicles, rip up the carpeting, put up posters, bring in tables and chairs that roll, and voila! They're then stumped when their shop is still low energy, no collaboration, and slow as all get out."

The point is, you can't just import artifacts or practices and expect them to have the same positive impact without the supporting cultural infrastructure.

In a different cultural environment, pair programming might lead to the "free-rider effect" where people don't pull their weight because there's someone else who will carry the pair. Or, in a different cultural environment, letting the team make promotion decisions could be a disaster if everyone is eager to promote everyone else in an implicit "I'll scratch your back if you scratch mine" agreement. In a different cultural environment, what an HTA learns about an end user's needs might be quickly dismissed by the project manager if the client disagrees.

So, yes, it's important to look closely at what Menlo does because they are doing amazing things. But it's also important to understand the deeper dynamics of their culture to understand why what they are doing works.

✦ ✦ ✦ ✦ ✦

As I mentioned at the outset of this Epilogue, I tried to capture some insights about culture in *The Thriving Organization: The Deep Dynamics of High-Performing Organizational Cultures*. There, I attempted to distill the lessons learned from my own experiences of nearly 30 years at W. L. Gore & Associates, as well as those learned from research into other, very different organizations, like Menlo. The goal was to identify some general characteristics of "good" cultures, characteristics independent of company age, size, industry, strategic intent, etc. The resulting model was derived from Schein's definition of culture, and particularly from his idea of various layers of cultural elements (*see* The Model of Thriving Organizational Culture in the appendix).

Although organizational artifacts and practices are often the shiny new objects that capture our attention, equally and indeed more important are what Schein calls the invisible (or tacit) dimensions of organizational culture. The "thriving organization" model of culture lays out three of these ever-deepening layers: *principles*, *values-in-use*, and *axioms* or fundamental assumptions. So, let's take a look at each of these in a bit more detail than what's been presented so far, and show how each is manifested in the Menlo culture.

PRINCIPLES

Principles are what directly guide action in organizations, and five principles seem to be generally operating in a variety of high-performing organizations:

1) Connect the dots.
2) Cultivate sensitivity to the outside world.
3) Enable collaborative emergence.
4) Invest in future (adaptive) capacity.
5) Expect leadership everywhere.

CONNECT THE DOTS.

People want to make sense of their world. They want to understand how what they do has an impact or makes a difference. With that in mind, it is typical for organizations to try to promote an understanding of their mission and vision, their strategy, and their purpose. Many organizations also hold quarterly or annual town halls where the CEO provides financial updates on how the organization is doing. Often, however, these efforts seem like little more than a set of talking points drummed up by someone in Corporate Communications. Nice to know, but so what? Even if employees find them informative, they rarely provide Jane or Joe with an understanding of what they are supposed to do on a daily basis. Now that I know this, what am I supposed to do differently? Anything? Or just talk a new game to show I'm onboard with this month's corporate lingo? How exactly do I contribute to the strategy? The mission? The financial results? What is our business model? How exactly do we make money? All too often, those dots aren't connected.

Even more problematic is that these attempts to communicate seem to follow the mantra of "talk to, talk to, talk to" rather than "talk with, talk with, talk with." That is, they are one-way presentations of information. What's also needed is an opportunity for people throughout the organization to tell their leaders what they're experiencing, what they're learning, what they're hearing from suppliers or customers. That is, the people who are experiencing the "dots" need to be able to ask the leaders, "What about this dot? What should I or we be doing about it?"

Too often, presentations on strategy or mission or values or purpose are at too high a level. They don't connect with what people are experiencing. And that lack of knowledge of what people on the ground are experiencing is a real danger in very hierarchical organizations. As one CEO put it, "The only thing worse than being in a tight place is being in a tight place and not knowing you're in a tight place."

How Menlo Innovations connects the dots. Menlo creates an information-rich, information-relevant environment. Menlo's mission statement creates a context for what information is relevant and provides guidance for meaningful communication. If you talk to people at Menlo, you find that to a surprising extent they actually think about their work in terms of "reducing suffering as it relates to technology...*today!*"

At a nitty gritty level, Daily Stand Up, the open access to the storycards posted on the Work Authorization Board, and the fact that at Menlo anyone can ask questions of anyone else, all this gives Menlonians a very good sense of what's going on, what's happening, project by project. People are not in the dark. Similarly, the Levels Poster makes compensation, including individual salaries, totally transparent. And the Financial Open Book Budget Management approach to company finances keeps everyone fully apprised of the company's performance. That everyone has access to viewing QuickBooks further inspires trust that there is nothing structurally keeping people in the dark.

There's no doubt that Menlo's small size makes it easier to connect the dots. But there are many small companies where folks are very limited in their understanding of how they actually contribute to the company mission, and where individual and company performance is pretty opaque. That's definitely not the case at Menlo.

CULTIVATE SENSITIVITY TO THE OUTSIDE WORLD.

Organizations exist to provide products or services to customers, and they continue to exist only to the extent that what they provide to the outside world costs less to deliver than the value it provides. That said, attentiveness to the outside world would seem to be an obvious requirement. But as Steve Denning and so many other writers have bemoaned, too many large and bureaucratic organizations have become too internally focused. The customer isn't the boss, my boss is the boss! That is, I work to keep my boss happy, not to keep the customer happy. Indeed, I often have no clue who my customer is, or how they are responding to what I do.

But cultivating a sensitivity to the outside world is more than just paying attention to current customers. For an organization to not get blindsided by external exigencies, the organization must take a broader view. What are we learning from *future* potential customers, even though they aren't customers now? Who are our competitors? What are they doing? What do our suppliers tell us about trends they see? What government regulations might be relevant to us in a year or two? What social, demographic, economic or technological trends will impact our business in the foreseeable future?

Organizations that do a good job of continuously assessing the outside world not only help themselves anticipate potential challenges and opportunities, they can also use the knowledge they gain as a source of value for their current customers.

How Menlo Innovations cultivates a sensitivity to the outside world. The Agile practice of getting working software in front of clients more frequently is one way of cultivating a better appreciation for client needs. And Menlo's weekly Show & Tells, with the client present, create a frequency of client feedback that keeps Menlonians sensitive to client wishes and needs. Perhaps even more critical is Menlo's use of High-Tech Anthropologists to research the wishes and needs of end users. It's worth noting that virtually all of Menlo's clients see the value in this practice, and they're willing to pay for it.

Both Show & Tells and HTA end-user research are baked into Menlo's sensitivity to the outside world. But Rich's public face to the outside world—his talks, his visits—also bring in an outside perspective on what's going on. So do the Liberty Project trade shows. So do, actually, all of the public visits to Menlo. Because the visitors are not just taking in information; by their questions and their comments, they are giving Menlonians a deeper appreciation for some of the challenges and opportunities they might be able to address.

ENABLE COLLABORATIVE EMERGENCE.

Collaboration is such a buzzword these days. It's almost as if whenever two people talk to each other, they are collaborating. But there's a big difference between communicating, cooperating, coordinating, and collaborating. These are all good things, and they should all be promoted in every organization! But when we collaborate, we are taking communication to a more impactful level. To truly collaborate, we have to bring together people with different perspectives in an environment of free and open discussion so that synergistically they can create something that no one of them would have come up with on their own. True collaboration occurs when the back-and-forth of debate and dialogue leads to an *emergent* solution. True collaboration is different from one person gathering up different ideas and perspectives and synthesizing them into a solution. And true collaboration is certainly much different from a group's brainstorming to come up with a unique solution, only to have the leader receive all the credit. True collaboration requires mutual respect, mutual appreciation, and mutual acknowledgment of contribution.

How Menlo Innovations enables collaborative emergence. Obviously, the Menlo practice of having everyone pair up enables collaboration. But it is striking how often individuals and pairs reach out to other individuals or pairs for additional input. Menlo seems to have institutionalized "micro-collaborations," quick interactions where one person says, "Here's what I'm thinking, can you help me improve it or come up with a better idea?" At Menlo people engage others not for approval but to open up the possibility that a different set of eyes will bring a different and potentially useful insight into an issue.

INVEST IN FUTURE (ADAPTIVE) CAPACITY.

Almost every organization invests for growth. They invest in research and development. They procure new and more efficient manufacturing equipment. They open up new offices and hire more salespeople. They have programs for professional development. And all of these actions are worth taking. But very often, organizations make these investments to build on what they already do. By contrast, when we invest in future adaptive capacity, we are investing to learn about things that may or may not be useful in the future. We are in some sense exploring through experimentation. Sometimes our exploration and experimentation are driven by what we see coming at us. Sometimes they are driven by capabilities we already have but can perhaps leverage in new ways. In either case we are extending beyond what we already know and venturing into the realm of what we don't know.

Christopher Worley, in his book *The Agile Corporation*, has explored the question of what makes it possible for some companies to regularly and over an extended period of time outperform their competitors. His conclusion is that, regardless of the performance of their market segment, the high performers in each segment allocate more time to learning than do lower performers. They may allocate less in a stable environment, and more in volatile environment, but in any case, high-performing companies avoid the temptation to focus only on harvesting the maximum they can from current capacities. No matter how strong they are currently, they invest in individual and organizational learning to increase future adaptive capacity.

How Menlo Innovations invests in future (adaptive) capacity. Menlo experiments. They try things. There is a low barrier to trying something new. If someone has an idea, they can run the experiment. I asked James Goebel if Menlonians had to clear their experiments with him or Rich or someone else before they ran them. James said that they didn't. He said he knew of only one or two experiments that were going on at that time. Then he did a "Hey Menlo!" and asked the room how many people were running experiments right then. More than a dozen hands shot up.

Take Menlo babies. When the first Menlo mom brought her baby to work, no one knew how the experiment would turn out. In fact, it worked out fine, and led to a series of upgrades and further experiments that led to Menlo being a pretty baby-friendly place for Menlo moms and Menlo dads, a fact that has paid off in terms of recruiting and retention. There was also Kealy's experiment of working remotely when she was temporarily providing care for an ailing parent. Menlo had always worked face-to-face. That was part of the process! But Kealy's successful experiment led to Randy's being able to work from Moscow on a permanent basis. And those experiments led to Menlo's being better positioned to deal with the coronavirus pandemic when people were forced to work from home.

EXPECT LEADERSHIP EVERYWHERE.

In most organizations, leaders have authority because of their positions. Leadership positions come with certain areas of responsibility and accompanying decision rights. Depending on their position, leaders may be able to hire, fire, write a big check, conduct a performance appraisal, give a raise, launch an initiative, develop departmental strategy, review reports from underlings, meet with higher-ups, and be the conduit for information flow up and down. But the general principle is that leaders set the agenda, pass that agenda down to their direct reports, who in turn pass it down to people on the front line. Everyone responds to their leader's agenda, and the leader then passes judgment on the adequacy of their response. As a result of the way information and directives cascade down through the hierarchy, the leaders at each level are the only people with direct contact with the leaders above them in the hierarchy, and as a consequence, they are the only ones with access to the organization's goals, priorities, and current levels of performance, all of which are crucial to the effective allocation of resources.

This command-and-control style of leadership has been the dominant conceptual model for decades, if not centuries. It clearly is being attenuated by environmental demands (from customers, suppliers, regulators, competitors) that require ever quicker responses from front-line staff. So, many organizations are now moving toward what is often called a "distributed" model of leadership, where decision making is much more widespread.

But even in these organizations, leaders are ultimately accountable for their unit's performance, and that gives them the authority to review and pass judgment on whatever initiative someone reporting to them might try. And unfortunately, even in organizations with a distributed leadership model, leaders sometimes use their authority to question, second-guess, or blame the people who report to them if something goes amiss. As a result, people learn to be very cautious about exercising any "leadership," and often run all decisions past their boss before taking them for fear of being called out for a mistake.

In high-performing organizational cultures, leadership is detached from position. Anyone can suggest, question, propose, critique, initiate. An individual's effectiveness as a leader—indeed their authority as a leader—will depend on, as the saying goes, whether anyone is willing to follow them. And that followership, in turn, depends not on position and title, but on the quality of the leader's ideas and the credibility the leader has built given prior experiences and track record.

Expect leadership everywhere is closely tied to the first principle of *connect the dots*. The capacity of employees to offer ideas, proposals, and initiatives that others will support—i.e. to exercise leadership—depends on their being well informed about what's going on at the 35,000-foot level as well as the ground level. They need *depth* of knowledge (that's what frontline workers typically bring), but they also need *breadth* of knowledge to sort out priorities, resource availability, and so on. In high-performing organizations, many if not most of the employees—and not just the people with titles—have this broader understanding of the organization and its work.

How Menlo Innovations expects leadership everywhere. Menlo decidedly expects leadership everywhere, and the cultural practices encourage and enable everyone to step forward as a leader. At the basic level of pairing, each pair partner is expected to offer suggestions, raise questions, offer critiques, try something different. Even if one pair partner is more senior and more experienced, the presumption is that the junior pair partner isn't just taking but offering whatever they can to contribute to the pair's success. Review of the pair's work isn't done by a boss, but by QA (to make sure things work as promised) and more importantly, by the client at each Show & Tell.

Another aspect of expecting leadership everywhere at Menlo is that teams do the hiring. Rich and James do not review those decisions. The teams set the pay-level, usually with input from others, but not necessarily from Rich and James. Termination, when it happens, is a team decision. When the sales revenues started to drop below a sustaining level, it was the team that pushed Rich and James to enfranchise a sales function that they had been loath to develop. (Of course, interestingly, the sales function that emerged was an example of collaborative emergence. It wasn't a cookie-cutter approach that many organizations might take: Let's go out and hire a few sales folks. It was a unique solution, crafted to fit the Menlo culture and the capabilities of Menlonians.)

VALUES-IN-USE

Almost every organization has some explicit statement about "Our Values," but all too often these values are more espoused than acted upon. (Enron famously listed "Integrity" as one of its values.) So, the key is: What values are actually promoted by everyday organizational action? While different organizations, including different thriving organizations, may have different values based on their unique strategic intents, these four values seem to cut across high-performing organizations, regardless of their differences:

1) Trustworthiness
2) Caring
3) Self-responsibility
4) Magnanimity

(Note: These are all values that individuals in the organization live by. They don't include organizational values like innovation or customer service or lean manufacturing that flow from the organization's strategic intent. As with individual values, these organizational values may either be just "espoused" or really accurate descriptors of how the organization proposes to compete.)

TRUSTWORTHINESS

Nearly everyone will agree that trust is a necessary element of a high-performing organization. But it's not like you can just ladle into your organization two heaping spoonfuls of trust. Trust isn't an input, it's an outcome. You get trust by being trustworthy. Being trustworthy generally means that you operate with good intentions, that you're competent at what you do, and that you act ethically. You deliver on what you say you're going to do.

How Menlo Innovations engenders trustworthiness. As with all of the individual values-in-use that we're looking at here, one of the key ways that Menlo sees to the trustworthiness of its employees is by screening for it in their hiring practices. The initial extreme vetting process allows Menlonians to get a quick read on an applicant's intentions. They can see if an applicant is driven by a "me-first" or "we-first" agenda. Later, the three-week engagement phase of the hiring process provides a deeper understanding of how, and whether, the applicant will fit into the Menlo culture, as well as a very granular sense of their competence and skill level.

Menlo's extended and in-depth hiring process also reveals an applicant's willingness and eagerness to become *more* competent. Are they willing to jump in and learn things that they don't know how to do? Because at Menlo, nobody achieves competence perfection. Everyone can learn new things, learn how to do things better, learn to improve old skills and develop new ones.

Another Menlo practice that helps support trustworthiness is the frequent question, do you have time for this? The point of the question isn't to cast doubt on a Menlonian's intention or competence. It's to remind everyone that saying you're going to do something is committing to an informal contract that everyone else will now expect you to live up to. It's easy in a low-fear culture where running the experiment is the norm for people to sometimes take on more than they can deliver. "Do you have the time for this?" is a way to support one another by hitting the pause button before someone makes a commitment they may not actually be able to deliver on.

CARING

For too many managers, and management theorists, the rational model of organizations precludes human emotion. Managers are supposed to make rational decisions and are not to be driven by feelings (although "intuitions"—a sort of supra-rationality—is sometimes praised). Indeed, emotions and feelings are often seen as antithetical to good organizational behavior. Leave your feelings at home! Don't bring them into the office!

But people are, in fact, emotional. They have feelings. Sometimes joyful ones and sometimes sad or distressed ones. And in high-performing organizational cultures, people are sensitive to the feelings of their colleagues. When their colleagues are feeling up, it feeds a positive cycle. But when their colleagues are feeling down, it elicits genuine concern. Can I do anything for you? Can I help? Want to talk about it? Caring is an expression of our full humanity—and it belongs at work.

Caring, maybe counterintuitively, extends to being willing to give others some tough feedback, if and when that seems necessary. If you really care for others, you don't want them to continue with behavior that is off-putting, or less than what they are capable of, or less than what is expected. Of course, when giving tough feedback, one can do it in a way that is respectful of the individual and coming from a place of good intention, rather than using "honesty" as a way to demean others and cut them down a notch.

How Menlo Innovations engenders caring. Menlonians often comment that Menlo is a place where you can be yourself at work. You don't have to pretend you're feeling okay when you're not. Not that everything devolves into a pity party. The work still needs to get done. But sometimes you get work done by addressing what's distressing rather than ignoring it. So, when people are sick, they're not expected to tough it out beyond a reasonable go. No, go home. When people are worried about how they're going to provide childcare for a newborn, they can relax and bring the little one to work. And when the little one is at work, there are suddenly plenty of willing "aunts" and "uncles" around to help out when need be. People look out for one another. If you're juggling work, commuting in with a spouse, and/or dealing with childcare, your pair partner will try to adjust his or her schedule to help ease your juggle.

For Menlonians, caring also extends to tough conversations they might need to have with one another, or even with a client. Jaron and his colleagues press Grant on his need to get comfortable with (and competent at) test-driven development. James is direct with Carl, the client of the Guinevere project, that one of the client's own developers might not be up to snuff and is hurting the project effort. When offered in a spirit of genuine helpfulness, such caring conversations lead not to defensiveness, but to positive change.

SELF-RESPONSIBILITY

Being responsible means doing what you're told. Being self-responsible means doing the right thing. It means when something falls apart, you don't try to deflect blame by saying, "I was just doing what I was told." People who are self-responsible are not into the blame game. Even though they may not be primarily responsible for a screw-up, they are always willing to consider whether there was something they did or said, or didn't do or didn't say, that led to the problem. Self-responsibility is at the heart of true problem-solving. It's about looking at a problem with your eyes wide open, being mindful of the big picture while digging down deep to ferret out the root causes—even if you've contributed to the difficulties.

How Menlo Innovations engenders self-responsibility. Even as they focus on their specific task at any given moment, to a remarkable extent Menlo people also seem to remain mindful of the big picture, expressed in their mission statement of "ending human suffering as it relates to technology." Menlonians operate within a problem-solving, or better—problem-avoiding—framework: How do we keep from creating suffering?

So, when the Quail software crashed and caused a serious problem for the client, the issue was never about blame. Michelle the PM immediately jumped into help mode, rather than waiting for the client to fix "their" problem. On the Guinevere project, the Menlo team and the client team agreed that both teams would run tests to locate whether a problem was showing up on the Menlo end or the client end. Once that was determined, only then would the decision about how to pay for the fix become relevant. And in the lunchtime conversation between Rich, James, and Pat, the threesome was trying to figure out a good path forward around a sticky problem, not to deflect potential blame, but to come up with a workable solution. They had a clear-eyed view of the problems on the client side, but that didn't prevent them from interrogating their own contributions to the situation. Similarly, with other projects and other problems, the questions are always: What did we contribute? What did others contribute? How do we resolve this in the best interests of all?

MAGNANIMITY

We all know folks who always seem to be looking out for #1, themselves. Maybe we succumb to that temptation ourselves on occasion. We want the credit for our work. We want the praise. We want the reward. It's all about me.

Magnanimity is the move from "me" to "we." When we operate with magnanimity, we know that whatever success we've had, others have had a hand in helping us along the way. And instead of claiming the credit for ourselves, we instead thank others for what they've contributed. Magnanimity is the appreciation at the heart of the insight that we accomplish more in partnership with others than by ourselves. We appreciate the different gifts and talents that others bring to our collective endeavors. Indeed, we are grateful for whatever successes others have because we all benefit from each individual's success.

How Menlo Innovations engenders magnanimity. It all starts with the extreme vetting interview. Applicants are paired and told their job today is to make sure their pair partner gets invited back for the second-round interview. What? You mean I'm not supposed to shine? Show I'm a star? I'm supposed to make someone else look good? If you can't get your head around this challenge, you're not likely to operate from a sense of magnanimity—and you're probably not a good fit for Menlo.

In too many organizations, if you win, I don't. At Menlo, success is a team effort. It's about what we collectively achieve. It's about the satisfaction and reward we get when we win—when we all win. With magnanimity, reward is not a zero-sum game.

✦ ✦ ✦ ✦ ✦

I hope it is rather clear that part of the reason the unusual Menlo practices work, like having everyone pair, or letting the teams decide on who to hire and how much to pay them, hangs on the way the values-in-use support the principles. If your employees are trustworthy, caring, self-responsible, and magnanimous, then you can safely share all financial data with them, trusting they will use that data to make good decisions. With people who live such values, you can better enable collaborative emergence. You can expect leadership everywhere.

Each of these values-in-use provides the supporting infrastructure that allows the principles to be productively manifested in Menlo's practices. Without the infrastructure of principles and values-in-use, Menlo practices could lead to disenchantment, rather than joy.

But now we need to dig a bit deeper to fully appreciate how the cultural infrastructure makes Menlo practices work.

AXIOMS

The values-in-use of trustworthiness, caring, self-responsibility, and magnanimity lead to what behavioral economists call "pro-social" behaviors. They lead to sharing and kindness and helpfulness and being able to take turns and play well with others. These are the behaviors that Menlo looks for in its extreme vetting process—to see if an applicant passes the "kindergarten test."

We'd like to think these values are fairly common in civilized society. Some behavioral economists have conducted large-scale studies that show, in fact, that only about 20% of Western-industrialized citizens are indeed "pro-social" almost all of the time. These they call "altruists." It turns out that about 30% of the citizens from these same societies are not much inclined to exhibit pro-social behavior. These they call "self-centered." For the remaining 50%, whether they behave in altruistic ways or whether they behave in self-centered ways depends on the perceived consequences of altruistic or self-centered behavior. In a variety of experimental games where outcomes can either be achieved collectively or individualistically, it turns out that the 50% will start out cooperatively. But if, during the game, self-centered players act in self-centered ways, that quickly shuts down the cooperative intentions of the 50%, who then start to play in win-lose mode.

What does all this have to do with values-in-use and high-performing organizations? Well, as Menlo shows, one way to get trustworthy, caring, self-responsible, and magnanimous behaviors is to hire people who value and actually demonstrate these behaviors! But as the above research suggests, for such people to continue to "play well with others"—for them to continue to display pro-social behavior—the environment must be set up to engender a pro-social orientation.

Here's a thought experiment that might make this clearer. Take a large population of people and measure their level of reluctance or willingness to tell the truth. Laid out on a graph, the results would probably be a normal distribution, with a few people always reluctant to tell the truth, no matter what, and a few people always willing to tell the truth, no matter what. But for most people, whether they would act in a fearful or fearless way about telling the truth in a given situation will depend on the circumstances. If they're at work and the boss is yelling, threatening, intimidating, and so forth, more people are likely to think that discretion is the better part of valor. They will move toward reluctance to tell the truth. They won't want to call attention to themselves by saying something the boss doesn't agree with or by making a mistake. On the other hand, if the boss is supportive, encouraging, the kind of boss who doesn't point fingers, then more people are likely to feel that they can take a risk, whether voicing an opinion or trying something that may not work.

In this example, the task for an organization isn't simply to hire people willing to always tell the truth (because they are few), but to create the conditions—the environment, the culture—where the population curve moves up the scale toward greater willingness to be truthful in all situations. It's not about getting 100% fearless folks. It's about creating conditions so more people can be less fearful. (This example, by the way, is what's behind work going all the way back to Edward Deming's focus on getting quality at work by driving out fear.)

So, what are the bedrock conditions that lead to less self-centered behavior and more pro-social behavior. More than assumptions, these axioms of high-performing organizational cultures must be there, must be true, to support the values-in-use and principles and practices of truly flourishing organizations. They are:

1) Treat individuals with respect.
2) Promote the common good.
3) Make decisions at the lowest level.
4) Focus on long-term success.

TREAT INDIVIDUALS WITH RESPECT.

Each person has an innate dignity that should be respected. Sadly, in too many organizations, the dignity of the individual is subordinated to a view of them as "human resources" to be leveraged by managers for the accomplishment of organizational goals. In such organizations, people are treated as expenses, not assets—despite the common refrain that "our people are our greatest asset." In short, they are treated as objects and not as people. Indeed, in many organizations, they are treated as easily replaceable objects.

Our me-first world often causes many of us to see others as falling into one of three categories: allies, adversaries, or irrelevant. That is, we see some folks as people whom we can call on to advance our agenda (and we try to figure out ways to persuade them—or to manipulate them—to do what is good for us). Other folks we see as adversaries, people who may potentially hinder or block our agenda (and we try to figure out ways to directly attack them or end-run them, often using our allies in our efforts). Finally, most people are essentially irrelevant to our agenda, and we don't give them much thought until it appears they might actually be potential allies or adversaries.

Treating people with respect requires that we view each individual as Real People having unique skills, talents, and interests. We value what they bring to the table now, and how they might learn and grow in the future. We see them holistically, not in terms of their resumes or positions or in terms of what they can do to—or for—us.

In high-performing organizational cultures, people experience themselves as being actors, not objects. They are people, not human resources. They experience the power of being able to make a difference, to contribute, to grow. They not only get to "bring their whole selves to work," they get to *become* their *best* selves at work.

How Menlo Innovations treats individuals with respect. Menlo starts treating individuals with respect before they're even hired.

Instead of looking over a job applicant's resume and making a yes/no decision based on education or past experience, candidates are invited in for the extreme vetting interview process so that Menlonians can get a sense of who these folks really are and how they think and whether they bring a me-first or a we-first orientation to work. Many, if not most, Menlonians actually do not have an academic background in computer science or computer engineering. Many, if not most, Menlonians who are High-Tech Anthropologists do not have an educational background in anthropology. What they had were the talents, skills, and passions to grow into those roles, and the values and attitudes that would allow them to be contributing members of Menlo.

Equally important at Menlo is the clear message that "You are more than a Menlonian." Lisa H. is not only a project manager. She is also the mother of a six-month old who's not yet ready for day care. Okay. Bring him to work. Kealy is not only a developer, but she is also the daughter of a parent that needed her at home to provide health care. Okay. Work from your parent's home. Because you're more than just what you do at work, you should go home at 5:00 or 5:30, and not work at night or on the weekend. Work when you're at work, but when you've put in your workday, go be the rest of who you are—family member, friend, neighbor, churchgoer, helper in the community. You are a Menlonian, but you're more than a Menlonian, and Menlo wants to honor—and encourage—that.

PROMOTE THE COMMON GOOD.

In most organizations, you will hear words like "One Team," and "We're all in this together." But all too often, the actions belie the words. In most organizations, for example, people are rewarded not for the achievement of organizational goals, but for the achievement of functional, department, or group goals. And so, instead of doing what's best for the company, I'll do what's best for my group. That, in turn, often leads to conflict over the resources—money, time, and people—that can make things happen. Instead of One Team, we're really many silos.

This siloed behavior isn't only a function of the reward system. In most organizations, other than perhaps being able to parrot back what the mission or the vision of the company might be, most people don't know how what they do actually contributes to that mission or vision. No one has actually connected the dots for them. So, lacking a clear understanding of the importance of the big picture, people sub-optimize and do what makes sense, given the narrow (siloed) perspective that they have.

In high-performing organizational cultures, the company mission is real, it's tangible, and people see a connection between what they do and the achievement of that mission. Information about company performance, challenges and opportunities, is widely distributed. At least to some extent, rewards are distributed on the basis of company performance, and not just on the basis of function or group/department/division performance. There is a strong sense of camaraderie as people genuinely root for each individual's success because that success will mean the company overall is more successful, which is good for all of us.

But in high-performing organizational cultures, the common good extends beyond the walls of the organization. High-performing organizational cultures work to promote the success of clients and customers, of vendors and suppliers, of communities. The common good is broader than what's in it for me.

How Menlo Innovations promotes the common good. Menlo's mission statement specifies the three stakeholder groups that they focus on: developers who want the code they write to make a productive contribution to the world, clients who want projects to be completed on time and on budget, and end users who want to be able to use technology products easily and without needing an advanced degree in computer science. When a project satisfies all of these stakeholders, it promotes the common good.

Because Menlo makes its financial information fully available to all its people, it is easy for Menlonians to understand how what they do not only impacts the mission, but also impacts the financial performance of their company. The mission requires profit because the company cannot be sustained without profit. For Menlo, profit ends up not being a goal or objective in itself. Rather, as Peter Drucker would have it, profit is a need. How much profit do we need to sustain the business we are in? Through their on-going discussion of the financial metrics they track, Menlonians grow in their understanding of the business and get to think like businesspeople, not just as developers or HTAs or PMs—as businesspeople looking out for the good of the entire business.

MAKE DECISIONS AT THE LOWEST LEVEL.

How do we make a good decision? Good decisions require knowledge depth and breadth. Depth refers to the granular knowledge that comes from on-the-ground experience and working every day on a task. Breadth refers to the broad knowledge of organizational context, priorities, and strategy. In most hierarchical organizations, breadth is the purview of managers, and depth is the purview of the doers. And in most hierarchical organizations, breadth is privileged in making all but routine decisions.

Here's an example: An engineer is working on a new product development project. She's sketched out a workplan with her boss and been given a certain budget and milestones for the project. As she moves to product testing, she discovers a new machine that might do a much better job of testing the prototypes than what is currently in her lab. She can't be sure. It will cost more than what was allocated in her budget. But it probably will shorten the testing phase. Should she buy the new testing machine? She understands the issue from a depth perspective. But she will probably go to her boss because the boss will know the breadth side of things. How crucial is this project to the organization? How tight will she (and he) be held to the budget? Will the possibility of cutting time-to-market be valued in this case, etc.?

There is an assumption in privileging breadth that depth knowledge is easily accessed and understood by higher-ups, while breadth knowledge is hard for organizational doers (like our engineer) to access and understand. Maybe one hundred years ago, this assumption would have been reasonable. But in today's organizations operating in conditions of uncertainty and rapid change, that may not be so true anymore. As organizations face greater and greater uncertainty, the breadth necessary to identify the consequences (and unintended consequences) of any action increases, and so decisions get kicked upstairs to someone with the requisite breadth to make the decision. At the same time, the speed of change is also increasing, so while decisions are traveling up and back down through the hierarchy, valuable time is lost. Conditions may change completely by the time a decision from the top is made. And lastly, there is the well-known effect in most organizations that information gets massaged as it goes up the ladder so as to make it more palatable and less likely to provoke blame.

For all of these reasons, and more, in high-performing organizational cultures it is increasingly the case that more and more consequential decisions are being made by the doers who experience the issue most directly. But this does not mean that such decisions are made only from deep knowledge and without a broader understanding. Rather, broad knowledge—priorities, strategies, organizational contexts—are more widely distributed, so that doers can understand their own work in the larger organizational context. This makes it easier for them to make certain decisions because they can apply both deep and broad knowledge. Having a breadth of knowledge also helps them "know what they don't know," and alerts them to when they need to find others who can help fill in the gaps in their knowledge before making a particular decision.

How Menlo Innovations makes decisions at the lowest level. Menlo enables widespread breadth of knowledge by literally putting the information out there. The posters that cover the walls, the Work Authorization Board that displays the status of every project, the Levels Poster, the running financial metrics written on a conference room wall, the weekly Open Books meeting—they all help ensure that every Menlonian, regardless of age, experience, role, or seniority, understands the big picture as well as the mundane details of the business. The fact that everyone is co-located, working in close proximity in an open space, (and using High Speed Voice Technology), also helps ensure that information flows easily and quickly to the person who needs it to make a decision. Additionally, Menlo's mantra of "make mistakes faster" gives Menlonians a certain tolerance level with making a decision, knowing they don't always have to be 100% right—as long as they learn from the consequences of a bad decision and share their learning with others.

FOCUS ON LONG-TERM SUCCESS.

It's a common complaint in many organizations: "We're too focused on quarterly numbers." Otherwise rational people sometimes do insanely stupid things when faced with an upcoming analyst call. Deals are cut to move revenues forward into this quarter—only creating more angst for next quarter. Investments are postponed. People are laid off. All to improve the quarterly numbers. *This* quarter's numbers.

In high-performing organizational cultures, the focus is on providing long-term value to customers and clients. It's a completely different lens. It's not that short-term results are unimportant. Sometimes you have to deal with a short-term crisis, or you won't have a long-term to worry about. But even in those situations, in high-performing organizational cultures, the short-term needs are still balanced with a consideration of the long-term.

Most telling about the difference between a short-term and a long-term focus is the organization's tolerance for, and encouragement of, learning. To some extent, every effort to learn means expending some amount of time, money, and human energy on something other than the task at hand. Learning is, to some extent, inefficient—at least in the short-run. And if you're only focused on the short-term, why bother to learn? Making an investment in organizational learning presumes that you are willing to bet that the future world will be different from the current world, future routines will be different from current routines, future issues will be different from current issues, and without some learning going on, the organization will be unable to cope or adapt (or exploit) those differences. You're willing to invest in learning because you want to be around for the future. Not all organizations make that investment, however.

Many are so focused on short-term efficiencies and short-term results that they do not adequately prepare themselves for a future that will certainly be uncertain.

How Menlo Innovations focuses on long-term success. Menlo invests in organizational learning, big time. The pair programming approach, with pairs changing every week, is a significant investment in the capacity of each individual Menlonian, but more importantly, in the organizational capacity of Menlo Innovations. While it might be more efficient (in the short-term) to hire programmers, say, with expertise in a particular programming language, and set them to work on all the projects that require or could use that language, Menlo chooses to up-skill all of their programmers with a large array of programming languages. By doing so, Menlo increases the flexibility it has in moving programmers to projects that require additional resources or moving them off projects that no longer require so many. Menlo decreases the negative unintended consequences of Towers of Knowledge who can't take vacations, can't get sick, and can't leave without all knowledge of a particular project or process being lost.

Menlo also tries new things—always with an eye to learning. Internal projects (not billable to a client) are often a way for Menlonians to learn a new programming language or try out a new app or read a new book. "Run the experiment" is a favorite Menlo expression, and its connotation is powerful. Unlike "launch a pilot project," run the experiment calls out that whatever happens—if the experiment works out as expected or not—what's important is that something is learned.

A particularly vivid example of Menlo's focus on long-term success is the Liberty Project, where the team is building an internal sales function. The project was born at a time of financial difficulties, when cash on hand and projects in the pipeline didn't appear sufficient to cover payroll for much longer. In some organizations, the solution would have been to lay some folks off. In other organizations, the solution would have been to hire a bunch of salespeople. Both solutions might have been more efficient, in the short-term. But what Menlo did instead was build an internal sales function, compatible with the Menlo culture, that draws upon the talents, skills, and interests of current Menlonians.

✦ ✦ ✦ ✦ ✦

What I hope has become obvious by now is that the Menlo culture is not a bunch of interesting but unrelated elements. Rather it is a rich network of interdependencies that create an organic whole. Rich Sheridan would say it's a "system." Ed Schein would say it's a "pattern."

So, the practice of pair programming most obviously promotes the principle of *enable collaborative emergence*. But it also promotes the principle of *invest in future (adaptive) capacity*. It depends on pairs being trustworthy, caring toward one another, self-responsible, and magnanimous in sharing credit. And those values are supported, and come to greater fullness, by axioms like *treat members with respect* and *make decisions at the lowest level*. It all has to work together if it's going to work at all.

In workshops where we've explored the Thriving Organizational Culture model, we often have small groups do an exercise where each person takes a turn at connecting one cultural element of the model with another, drawing a line on the culture map between the two elements, and speaking out loud the nature of the connection they see. People first make the obvious connections. *Connect the dots* makes it possible for the organization to *make decisions at the lowest level*. And *trustworthiness* allows the organization to share the information that allows people to connect the dots, and therefore permits *making decisions at the lowest level.* But then, as the group continues making these connections, invariably a light bulb will go off for someone who will say, "I think all the elements are actually connected to all other elements!" And that's what it means to say that a culture is "a rich network of interdependencies that create an organic whole."

But you don't get the whole all at once. It is, after all, organic. It grows. It evolves. Rich Sheridan likes to quote Gall's Law, "A complex system that works is invariably found to have evolved from a simple system that worked. A complex system designed from scratch never works and cannot be patched up to make it work. You have to start over with a working simple system."

✦ ✦ ✦ ✦ ✦

No organization is perfect and Menlonians will be the first to say that Menlo is still a work in progress. Yet they have, over the years, worked from a particular vision of what an organization can and should be. One that people love. One that brings joy. One that promotes true human flourishing. That vision has allowed them to "run lots of experiments" and organically add practices that "make sense" given the foundational culture that they have.

I hope this portrait of a day in the life of a thriving organization, along with some ideas about what the culture of such an organization entails, will help you build such a culture in your own organization. In doing so, look first at which Menlo practices already line up with your existing culture pattern/system. Leverage what you already have. Start small, maybe just where you are and where you have influence. Try one thing. Run the experiment. See what works. Go from there. Persevere! And always, always, seek the joy!

ABOUT THE AUTHORS

Michael Pacanowsky is the former Gore-Giovale Chair in Business Innovation in the Bill and Vieve Gore School of Business at Westminster College in Salt Lake City, Utah. He is the Founding Director of the Center for Innovative Cultures, whose mission is to "help organizations thrive, by unleashing the talent, passion, and potential of people at work." He is the author of *The Thriving Organization: An Exploration into the Deep Dynamics of High-Performing Organizational Cultures.*

Susan Arsht is an assistant professor of management in the Bill and Vieve Gore School of Business at Westminster College and former Executive Director of the Center for Innovative Cultures. She teaches courses in international business, citizen diplomacy, non-profit organization and business communication.

Vicki Whiting is a professor of management in the Bill and Vieve Gore School of Business. She teaches courses in leadership and organizational behavior. She is the author of *In Pain We Trust: A Conversation Between Mother and Son on the Journey from Sickness to Health.*

Sara D'Agostino recently graduated with a Master of Science in Finance from the University of Utah and is currently working at Scalar as a Valuation Associate.

Maggie Fischer is a Collateral Management analyst in the Corporate Treasury division of Goldman Sachs in the Salt Lake City office.

Rachel Iverson spent a year traveling and teaching English as a second language online in Vietnam. She now teaches English, history, and AP Psychology at Juan Diego Catholic High School in Draper,

Utah, her own high school *alma mater*.

Elizabeth Johnson is the Marketing and Events Manager for the Economic Development Corporation of Utah.

Cole Polychronis is pursuing a Ph.D. in Computing at the University of Utah with a focus on crisis informatics and how people use technology and spread information in times of natural disaster and political crisis.

At the time of the research for this book, Sara, Maggie, Rachel, Liz, and Cole were all Honors College students at Westminster College.

Jim McGovern is a writer living in Cambridge, Massachusetts. He previously assisted in the writing of *The Thriving Organization*.

APPENDIX

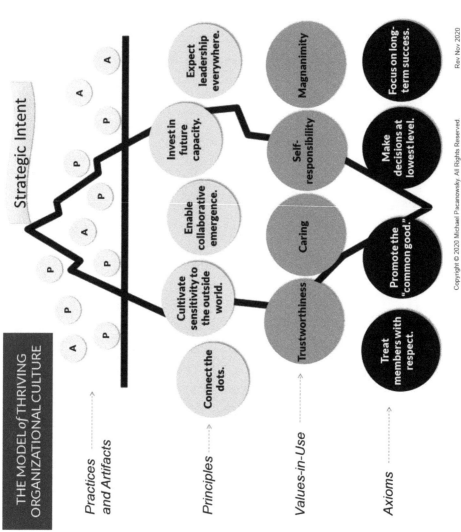

CPSIA information can be obtained
at www.ICGtesting.com
Printed in the USA
BVHW090931271021
619822BV00003B/14